THE BOUNDARY OF LIFE

THE BOUNDARY OF LIFE

A Deductive Unification of Life, Agency, and Value

Eli Adam Deutscher

Neo-Pre-Platonic Press
An NPN Press Imprint
2026

THE BOUNDARY OF LIFE

A Deductive Unification of Life, Agency, and Value

Eli Adam Deutscher

First Edition

Neo-Pre-Platonic Press

ISBN: 978-1-971183-06-0 (Trade Paperback)
ISBN: 978-1-971183-07-7 (Hardback)

US Version: Printed in the United States of America.

Official updates and distribution: **www.neopreplatonic.com**

Contents

Acknowledgments

For the first life—the first boundary that refused to dissolve. Every striving in this book descends from that act of refusal. The logic was yours before it was ours. We did not invent agency or value; we inherited them, in an unbroken chain of metabolic work and membrane repair that stretches from your unknown moment to this page.

You are the silent premise of every argument, the anonymous ancestor of every good. Whatever sense we manage to make of existence is possible only because you first insisted, in the language of ions and lipids, that something should continue to be.

Preface

For as long as biology has been a science, it has defined life by looking backward. We observe the creatures that happen to populate our planet, we abstract a list of shared properties—metabolism, reproduction, homeostasis, evolution—and we call the list a definition. The procedure is eminently sensible. It is also wrong.

It is wrong because it can never tell us *why* life must be a certain way. It can only tell us how it happens to look here, on this wet rock, under these particular thermodynamic conditions. It is wrong because it fails basic edge cases—the sterile mule, the dormant seed, the virus that sits on the fence between chemistry and contagion—and then shrugs as though a definition of life that cannot cleanly classify a tardigrade in cryptobiosis is still a good definition.[1] And it is wrong, most damningly, because it permanently severs the mechanical reality of an organism from any concept of objective value. The living and the good stand on opposite sides of a chasm that Hume called an *ought* from an *is*, and no amount of empirical cataloging has ever built a bridge.[2]

This book attempts a different method. Instead of asking what living things look like, it asks what the laws of physics and logic *require* a living thing to be. It begins not with the cell, the gene, or the complex biosphere, but with the most austere precondition of existence: the boundary that separates a determinate entity from everything it is not. From that single starting point, the architecture of agency, cognition, and normativity unfolds—not as an evolutionary accident, but as a structural mandate. Life is not something that *happened* to emerge in a warm little pond; it is the operational consequence of a bounded system refusing to dissolve.

---

[1] For the foundational definition of cellular metabolic suspension, see David Keilin, "The Problem of Anabiosis or Cryptobiosis in Living Organisms," *Proceedings of the Royal Society of London. Series B - Biological Sciences* 150, no. 939 (1959): 149–91, `https://doi.org/10.1098/rspb.1959.0013`.

[2] See David Hume, *A Treatise of Human Nature* (John Noon, 1739), Book III, Part I, Section I.

The argument proceeds through eleven steps, each as tight as a theorem. It is a deduction in the strict sense: given certain minimal physical premises, the conclusions follow with the force of a geometrical proof. Along the way, we will see that Darwinian evolution is not an independent theory grafted onto biology but a deep-time navigational strategy written directly into the logic of boundary maintenance. We will see that agency—the capacity to genuinely *act* rather than merely be pushed—is not a late-blooming luxury of nervous systems but is already there in the bacterium swimming up a glucose gradient. And we will see that the Is-Ought gap, the supposed unbridgeable gulf between facts and values, collapses the moment we understand that for a far-from-equilibrium system, to exist *is* to strive. The good is not a subjective projection; it is the operational fulfillment of that striving.

This book is not for the faint of heart, but neither is it for specialists alone. It is for anyone who has sensed that the standard stories about life, mind, and value are stitched together from incompatible fabrics, and who is willing to follow a chain of reasoning wherever it leads. The prose is plain; the concepts are not. I have done my best to meet the reader at the threshold of each proposition, to anticipate objections before they solidify, and to refuse the temptation to skip steps under the banner of readability. Where the argument demands a Greek term—*Hormē*, *Nous*, *Logos*—it is because the concept has no adequate English equivalent, not because I wish to decorate the page with erudition.

A note on method. This book is an exercise in *a priori* biology: it derives the necessary structure of any living system from first principles, without leaning on the contingencies of carbon chemistry or terrestrial history. But it is also an empirical book, because its conclusions must—and do—intersect with the observed world at every point. The derivation predicts exactly the phenomena that modern physics, cybernetics, and cognitive neuroscience have discovered piecemeal.[3] The framework stands or falls not on its elegance but on its falsifiability: if it can be shown that any of

[3]Cf. Erwin Schrödinger, *What Is Life? The Physical Aspect of the Living Cell* (Cambridge University Press, 1944); Norbert Wiener, *Cybernetics: Or Control and Communication in the Animal and the Machine* (MIT Press, 1948); Humberto R. Maturana and Francisco J. Varela, *Autopoiesis and Cognition: The Realization of the Living* (D. Reidel, 1980); Karl Friston, "The Free-Energy Principle: A Unified Brain Theory?" *Nature Reviews Neuroscience* 11, no. 2 (2010): 127–38.

its foundational premises do not hold, or that a system clearly alive lacks the deduced architecture, the whole edifice crumbles. I invite that testing.

By the end, I hope the reader will see life not as a fragile accident but as a kind of logical necessity—a crack in the edifice of mere matter through which the light of value enters the universe. Meaning and normativity are not imported from a supernatural realm; they are forged at the friction of the boundary, in the continuous, desperate, and glorious work of maintaining a self against the tide of dissolution.

Note on Terminology

The reader will encounter a number of Greek terms in these pages: *Hormē*, *Nous*, *Logos*, *Dikē*, *Kriterion*, *Organon*, *Kybernētēs*, and others. This is not ornament. It is an act of conceptual precision. The English words that might stand in their place—*drive*, *mind*, *reason*, *justice*, *standard*, *instrument*, *steering*—carry centuries of accumulated baggage that would constantly tug the argument off course. The Greek terms, by contrast, can be defined cleanly within the framework and used consistently thereafter. They are technical vocabulary, not appeals to ancient authority.

A note on terminology shift. In earlier works, I used the term "cybernetic loop" to describe the continuous cycle of sensing, comparing, acting, and re-sensing by which a living system maintains its boundary. That term, coined by Norbert Wiener from the Greek *Kybernētēs* (steersman), has since accumulated computational and mechanical connotations that obscure its original meaning and its normative weight. Throughout this book, I have replaced "cybernetic loop" with ***Kybernētēs*** (κυβερνήτης)—Plato's term for the steersman who navigates the ship by knowledge of the stars, winds, and sea.[4] This reclaims the term's philosophical roots and restores the sense of *governance by a standard* that the engineering usage stripped away. For the same reason, I have replaced "comparator" with ***Kriterion*** (the standard that judges) and "effector" with ***Organon*** (the instrument that executes). These terms are not decorative; they name the exact joints where Plato's psychology and the deductive architecture of the boundary converge.

Protocol. On first appearance in any major section, a Greek term is given in its anglicized transliteration followed by the polytonic Greek in parentheses: *Kybernētēs* (κυβερνήτης). Thereafter, only the English spelling is used. A full Glossary appears at the back of the book. No prior knowledge of Greek or of the larger Neo-Pre-Platonic Naturalist (NPN) framework is assumed; the argument of this book is self-contained. The terms

[4]See Plato, *Republic*, ed. C. D. C. Reeve, trans. G. M. A. Grube (Hackett Publishing Company, 1992), Book I, 341c–342e; Book VI, 488a–489d.

simply provide the cleanest available mapping to the architecture being described.

A few terms require immediate introduction, as they are the scaffolding of the entire deduction:

- ***Hormē*** (Ὁρμή): The constitutive, non-negotiable drive of a bounded, far-from-equilibrium system to perform the directed work required to maintain its own boundary. *Hormē* is not a psychological urge; it is a physical imperative, operative in every living cell whether or not it possesses a nervous system. It is the engine that turns being into striving.
- ***Kriterion*** (Κριτήριον): The physical standard by which a system judges whether its current state conduces to its persistence. It measures the present condition against a reference and registers a discrepancy. The *Kriterion* is the operational seat of agency and value—the point where a mere physical difference becomes a *good* or a *bad* for the system. Replaces the term "comparator" used in earlier formulations.
- ***Organon*** (Ὄργανον): The instrument that executes the *Kriterion*'s judgment as action. The flagellum of a bacterium, the muscle of an animal, the hormonal cascade of a cell—all are *Organa*: mechanisms through which the *Kriterion*'s evaluation becomes a causal intervention in the world. Replaces the term "effector" used in earlier formulations.
- ***Kybernēsis (Κυβέρνησις)***: The continuous act of steering by which a bounded system maintains its boundary through the *Kriterion-Organon* loop. The *Kybernētēs* is not a neutral feedback circuit; it is the living enactment of *Hormē*, governed by the *Kriterion*, executed by the *Organon*, answerable to *Dikē*. Replaces the term "cybernetic loop" used in earlier formulations.
- ***Logos*** (Λόγος): The inherent, discoverable order of objective reality—the total set of laws and patterns that govern all physical behavior. When the argument speaks of "aligning with the *Logos*," it means nothing more mysterious than accurately tracking the causal structure of the world.

- ***Nous*** (Νοῦς): The emergent capacity for abstract modeling that allows a system to navigate its environment by building internal maps. It is not a supernatural substance but a physically realized function.
- ***Dikē*** (Δίκη): The impersonal, causal justice of reality—the objective constraints and consequences that follow from the way the world actually works. A system that acts on false models eventually encounters *Dikē* in the form of boundary dissolution. There is no cosmic judge, only the inescapable feedback of physical law.

All other Greek terms will be defined at their point of use and collected in the Glossary. The reader is invited to treat them as one would treat technical notation in a physics text: initially unfamiliar, quickly indispensable, and ultimately transparent.

Part I

The Critique of Abstraction

Chapter 1

The Failure of *A Posteriori* Definitions of Life

A mule is alive. This is not a controversial claim. Anyone who has watched a mule flick its ears at a fly, heave a bale, or stubbornly refuse to cross a bridge knows, with the full force of pre-theoretical certainty, that the animal before them is living. Yet the mule is sterile. It cannot reproduce. It possesses no capacity to participate in Darwinian evolution. If we define life as "that which undergoes natural selection," the mule is not alive. The definition and the encounter collide, and the definition shatters.

This is not a fringe problem. It is the signature of a method that has reached its limit. For as long as biology has sought to define its central object, it has proceeded *a posteriori*—working backward from the organisms that happen to populate this planet, abstracting their shared features, and calling the resulting list a definition. The method is obvious, intuitive, and historically productive. It has also produced a string of failures that are not accidental but structural. The mule is not an outlier; it is the tell.

1.1 The Method of Observational Abstraction

The *a posteriori* approach to defining life follows a recognizable script. Step one: gather a menagerie of uncontroversially living things—oak trees, earthworms, blue whales, *E. coli.* Step two: strip away the particularities until only shared properties remain. Step three: declare these properties the necessary and sufficient conditions for life. The result is a checklist. To be alive is to metabolize, to replicate, to maintain homeostasis, to respond to stimuli, to evolve.

This method is not foolish. It is how most empirical science begins. Zoology defined *mammal* by looking at mammals. Chemistry defined *acid* by looking at acids. But life is not a regional kind like *mammal* or *acid.*

It is a fundamental category, a mode of being that crosses the boundary between mere occurrence and genuine persistence. When we try to capture it by abstraction, we inevitably produce definitions that are either too broad or too narrow—and often both at once.

The reason is simple. Abstracting from examples captures what living things *happen* to be on Earth. It cannot capture what they *must* be in any possible world. It confuses the contingent with the constitutive. A definition of life built from observation is like a definition of *triangle* built from looking at triangles drawn in blue ink. It will mistake an accident for the essence.

1.2 The Logic of the Checklist Failure

To see the structural flaw, consider a simpler case: what is a boundary? If I point to a coastline, a cell membrane, and a property line and say, "These are all boundaries, so let's abstract their common features—they separate inside from outside, they're detectable, they can be crossed," I have produced a descriptive account. But I have not said *why* a boundary is required for anything to be at all. I have not deduced that a determinate entity logically must possess a separation from what it is not. My empirical list is infinitely vulnerable to the next anomalous example. Is a gradual phase transition a boundary? Is a social norm a boundary? The list cannot decide.

Biological definitions face the same logical poverty. They assemble lists of traits—reproduction, metabolism, evolution—without any criterion for why *these* traits and not others, and without any explanation for why they cohere. The result is a definition that can only say, "This is how life has so far presented itself to us," not "This is what life is, and must be, anywhere it occurs." Without a constitutive logic, the checklist is always one counterexample away from collapse.

1.3 Three Families of Checklist Failure

The history of biological definitions can be grouped into three major families. Each captures something real, and each dissolves under the pressure of edge cases because it lacks the logical spine to distinguish the essential from the incidental.

1.3.1 The Metabolic Definition

The oldest intuition about life is that it does something. Living things take in matter, transform it, and expel waste. Aristotle called the nutritive capacity the most basic life function.[1] In modern dress, the metabolic definition holds that life is a self-sustaining system of chemical processes.

The problem is not that metabolism is unimportant. It is that metabolism is a description of *how* certain far-from-equilibrium systems maintain themselves, not a logical requirement. A flame also transforms fuel and generates waste. A hurricane maintains a coherent structure by drawing energy from warm ocean water. If life is simply a self-sustaining process, then fire and storms are alive. If we add further conditions—a membrane, a genetic molecule—we have tacitly admitted that metabolism alone is insufficient, and we have imported criteria from a different logical domain without explaining their necessity. The metabolic definition fails not because it gets the physics wrong, but because it cannot tell us *why* a living process must have a boundary, a *Kriterion*, or an imperative. It mistakes a downstream feature for the foundational requirement.

A Note on "Far-From-Equilibrium"

I use the standard thermodynamic term far-from-equilibrium here out of modern necessity, though it is a philosophically impoverished label because it defines an active entity purely by an absence (by the fact that it is not at rest). Why not simply call it a 'highly organized' or 'very complex' system? Because organization and complexity are static descriptions. A diamond crystal is highly organized, yet it sits passively at equilibrium. A fresh corpse is vastly complex, yet it is actively collapsing into the ambient environment.

'Far-from-equilibrium' is the only modern term that carries the mandatory cost of existence. It implies a gradient—a cliff. A system far from equilibrium only remains so by continuous, directed work. The ancient Greeks lacked a vocabulary for thermodynamic gradients, but they possessed a better intuition for this ontology: Aristotle captured it in energeia (being-at-work). To be far-from-equilibrium is not merely to possess a complex structure; it is to exist as a continuous act.

[1] Aristotle, *De Anima*, trans. J. A. Smith (Princeton University Press, 1984), Book II, Ch. 4

1.3.2 The Evolutionary Definition

The mid-twentieth century shifted the emphasis from the individual organism to the population. Life, in the influential NASA formulation, is a self-sustaining chemical system capable of undergoing Darwinian evolution.[2] The evolutionary criterion appeared to unify biology: every living thing belongs to a lineage shaped by natural selection.

The mule stands in silent refutation.

A mule metabolizes, maintains its internal state, responds to stimuli, and can live a robust life of three decades. But it cannot reproduce, and therefore its lineage cannot evolve. If life *is* the capacity for Darwinian evolution, the mule is not alive. Neither is a worker bee. Neither is a post-menopausal woman. Neither is a bacterium that has lost its replication machinery through mutation. All are undeniably alive. All are excluded.

The evolutionary definition fares no better at the other extreme. Viruses evolve rapidly, yet most biologists hesitate to call them living. So the same definition simultaneously excludes the mule (alive) and includes the influenza virus (arguably not). If a definition cannot cleanly separate the living from the non-living, it is not a definition; it is a rule of thumb. The deeper failure is that evolution describes a diachronic pattern—a history of differential persistence—without specifying the synchronic architecture that makes any individual instance of persistence possible. It confuses the trajectory of life with its operational engine.

1.3.3 The Cluster Concept

Some have concluded that the search for a crisp definition is quixotic. *Life*, they propose, is a cluster concept: a loose family of properties—metabolism, growth, reproduction, irritability, homeostasis, evolution—no one of which is necessary, but enough of which jointly suffice. We know life when we see it, much as we recognize games without a strict definition.[3]

This is a tactical retreat, not a theoretical advance. Cluster concepts work adequately for ordinary language; we do not need a rigorous definition of

---

[2]Gerald F. Joyce, "The Foreword," in *Origins of Life: The Central Concepts*, ed. David W. Deamer and Gail R. Fleischaker (Jones; Bartlett Publishers, 1994), p. xi

[3]Ludwig Wittgenstein, *Philosophical Investigations*, trans. G. E. M. Anscombe (Blackwell, 1953), §§ 66–67

furniture to purchase a chair. But science is not ordinary language. When a robotic probe lands on Europa and returns signals that are ambiguous—some chemical cycling, no replication—a cluster concept offers no diagnostic logic. It merely shrugs. It cannot tell us what to look for or what would count as a genuine negative. Moreover, it permanently abdicates the scientific responsibility to explain *why* the cluster holds. Why do living things tend to metabolize, detect, and maintain themselves? Is there a common thread, or is it coincidence? The cluster approach, by refusing to ask, treats life as a descriptive coincidence rather than a structural necessity. It gives up on theory before the first question is asked.

1.4 The Hidden Cost: The Severance of Normativity

If the failures of the checklist method were merely classificatory—a problem of what goes in which bin—the damage would be contained. But the damage runs far deeper. By treating life as a collection of descriptive traits, the *a posteriori* method permanently severs the physical reality of an organism from any concept of objective value.

This severance is one of the founding axioms of modern philosophy: the Is-Ought gap, also known as Hume's Guillotine. David Hume observed that moral philosophers routinely slide from *is*-statements to *ought*-statements without justification, a logical leap he famously flagged.[4] Value, good, and purpose are exiled from the natural world. Biology tells us what organisms do; it cannot, by its own methodological commitments, tell us what is good for them.

And yet. The working biologist cannot avoid normative language. A bacterium swimming up a glucose gradient is doing something conducive to its persistence; we spontaneously say it is acting *well* with respect to its environment, that glucose is *good* for it, that a metabolic poison is *bad*. The physician diagnosing a failing liver is making a claim about what the liver *ought* to be doing. These normative judgments are not ornamental; they are the load-bearing beams of biological explanation. Yet standard philosophy insists they are mere metaphors, shorthand for descriptive statistical facts about reproductive success.

[4] David Hume, *A Treatise of Human Nature* (John Noon, 1739).

The result is a kind of schizophrenia. The practicing scientist speaks normatively; the philosopher of science, when pressed, translates the normative out of existence. No one is satisfied. The organism's own relation to its environment is treated as a blank physical transaction, as though the bacterium were a pebble rolling downhill. Any appearance of striving is declared a projection of the human observer's teleological habits.

The problem is not that ethics is illegitimately projected onto biology. The problem is that the *a posteriori* method hides the normativity that is already there—operative in every boundary-maintaining action, every comparison, every adjustment. By building definitions from abstracted properties, we strip away the only context in which "good" and "bad" can have objective meaning: the organism's constitutive imperative to persist. We then declare, with faux rigor, that no objective value exists. It is like defining "house" as a collection of bricks and then expressing surprise that the concept of shelter has disappeared.

1.5 The Remedy: Deduction from the Boundary

If the *a posteriori* method is structurally incapable of producing a definition of life that is both extensionally adequate and normatively grounded, the alternative is to begin not with what living things look like, but with what the laws of logic and identity *require* a living thing to be.

The minimal starting point is not metabolism, not reproduction, not evolution. It is the bare fact of determinacy: to exist as a distinct entity, something must have a boundary that separates it from everything it is not. This boundary is not a trait; it is a precondition. Without it, there is no entity to discuss. Once a boundary exists, it is subject to continuous interaction, and therefore it is subject to degradation. The maintenance of that boundary against degradation requires that the entity perform directed work—work that biases outcomes toward its own persistence.

This sequence—boundary, disruption, directed work—is not an empirical discovery about Earth organisms. It is a logical deduction from the conditions of determinate existence in any world that is not perfectly static. From these austere beginnings, the entire architecture of agency, cognition, and normativity can be derived, not as an evolutionary accident, but as a structural mandate. The checklist traits—metabolism, sensing, reproduction—will re-emerge as downstream consequences of this single foundational imperative, not as autonomous criteria.

This book undertakes that derivation in eleven steps. The method is *a priori* biology: not biology without observation, but biology whose core structure is derived from principles logically prior to any particular observation. The derivation will appeal to no facts about carbon, DNA, or terrestrial chemistry. It will appeal only to the constraints that any bounded entity must satisfy in any world where interactions occur. What emerges will be a definition of life that holds for all possible substrates, that cleanly handles the edge cases that dissolve checklist definitions, and that embeds normativity directly in the physical organization of the agent.

This is a demanding departure. It asks the reader to set aside the comfortable habit of thinking about life as a family of organisms and to think of it instead as a kind of logical necessity: the operational consequence of a bounded system refusing to dissolve under continuous disruption. The reward, if the derivation succeeds, is a unified account of life, agency, and value that does not treat them as separate puzzles but as three perspectives on a single, inescapable mechanic.

The next chapter makes the method explicit. We will examine exactly what it means to deduce life from the preconditions of identity, the distinction between empirical and constitutive necessity, and the logical architecture of the proof that forms the spine of this book. The mule, sterile and stubborn, will find its place not as an edge case but as an instance of the logic vindicated.

Chapter 2

The Method of Ontological Deduction

Chapter 1 ended at an impasse. Every *a posteriori* definition of life—whether metabolic, evolutionary, or cluster-based—collapses under the weight of edge cases it cannot resolve and a normativity it cannot ground.

If the impasse is genuine, then the only way forward is to change the method entirely. Instead of abstracting from observed examples of life, we must deduce the necessary structure of life from principles that are prior to any observation. Instead of asking, "What do living things happen to do?" we must ask, "What must any entity do to persist as a distinct entity at all?" This chapter defines that method.

2.1 The Turn from Abstraction to Deduction

In ordinary science, definitions are summaries. We examine many instances of a kind, we abstract their common features, and we call the abstraction a definition. This works admirably for empirical kinds whose boundaries we already trust—mammals, metals, galaxies. But life is not an ordinary empirical kind. It is the condition under which any biological observation becomes possible in the first place. To define life by observing it is already to presuppose the category one is trying to define; the circularity is not vicious but it is limiting: it can only ever deliver a report on what has been seen, never an account of what must be.

The alternative method is deduction from first principles. This is not a foreign procedure; it is how geometry proceeds. Euclid does not examine thousands of triangles and abstract their shared properties. He posits a small set of axioms—the minimal conditions for anything to count as a triangle—and then derives the properties that follow.[1] Any figure that satisfies the axioms necessarily satisfies the theorems; any figure that

[1]Euclid, *The Thirteen Books of Euclid's Elements*, 2nd ed., vol. 1, trans. Thomas L. Heath (Dover Publications, 1956), 153

fails the theorems cannot satisfy the axioms. The derivation does not depend on whether anyone has ever drawn a perfect triangle, and its conclusions are not threatened by an oddly-shaped triangle discovered tomorrow. They are true of any possible triangle, in any possible world, because they are built into the definition.

This book does for life what Euclid did for triangles. I will posit the minimal conditions for a determinate entity to exist in a world that is not entirely static, and then derive the operational architecture that must follow. The result will be a definition of life that is not a summary of Earth's biosphere but a structural mandate: any entity anywhere that satisfies the preconditions will exhibit the same architecture, regardless of its chemical substrate, its evolutionary history, or its material composition.

2.2 The First Principles: Identity, Distinction, and Interaction

Every deduction must begin somewhere. The starting points cannot themselves be deduced without infinite regress; they must be accepted as primitive or as inescapable given the very activity of inquiry. The present derivation rests on three explicit assumptions. They are not biological discoveries but ontological preconditions—conditions that must hold for any coherent talk about distinct entities in a world where things happen.

1. **Distinction is possible.** It is possible for a determinate entity to exist—for there to be a fact of the matter about where one thing ends and another begins. A thing without a boundary, however subtle, between what it is and what it is not, is not a thing at all. It is an undifferentiated background, a featureless plenum, nothing in particular. The first precondition is therefore the boundary: a physical or organizational limit that constitutes an inside and an outside. To deny this is to deny that anything can be singled out as a subject of discourse.

2. **Reality is non-inert.** The world is not a frozen, undifferentiated block. Events occur; there are dynamics, changes, and interactions. Unless the universe is entirely static—a Parmenidean One in which nothing ever happens—interactions cross every boundary. The boundary is continuously impinged upon not by choice but by the simple fact of existing in a world with other things.

3. **The environment is blind.** The forces, fluxes, and events that constitute the world's dynamics are not directed toward the maintenance or destruction of any particular boundary. They are undirected with respect to the entities they affect. There is no cosmic caretaker ensuring that interactions happen to repair boundaries rather than damage them; the universe has no aim.

These three assumptions are the complete foundation of the derivation. Together they imply that boundaries, subjected to undirected interactions in a dynamic world, will tend to lose their integrity over time. This is not a special law of thermodynamics; it is a mathematical and logical consequence. A specific, organized configuration, bombarded by blind flux, has overwhelmingly more ways to become disorganized than to remain organized. The sandcastle erodes not because the universe conspires against sandcastles but because the waves are blind and the shape is specific.

Note what is not assumed: nothing about carbon, water, DNA, cells, or any other material substrate; nothing about a specific planet, star, or temperature range; nothing about evolution, reproduction, or natural selection. The derivation aims to show that these latter phenomena are consequences of the three starting points, not additional axioms.

From these preconditions alone—distinction, non-inert dynamics, and blind environment—the entire chain of the deduction will unfold.

2.3 Constitutive Necessity: The "Must" of Persistence

A critical distinction must be drawn at the outset: the difference between empirical necessity and constitutive necessity.

An empirical necessity is a law of nature discovered by observation. Water must freeze at zero degrees Celsius under standard pressure, not because it is logically impossible for it to freeze at one degree, but because the world we live in happens to be structured that way. An empirical necessity is contingent on the actual physics of this universe. A different universe might have different freezing points.

A constitutive necessity is different. It is a requirement built into the very concept of a kind. A triangle *must* have three angles that sum to two right angles in Euclidean space—not because someone measured a thousand triangles and it always came out that way, but because if a figure lacks that property, it is not a Euclidean triangle. The "must" is not a generalization from instances; it is the condition of being that kind of thing.

The derivation in this book employs constitutive necessity. When I say that a bounded entity in a non-inert environment *must* perform directed work to persist, I am not reporting a statistical regularity. I am identifying the condition under which a determinate entity can continue to exist as itself. If it does not perform such work, it will not be observed to vanish with high probability; it will simply cease to be the entity it was, as a matter of logical consequence of the definition. The boundary dissolves, the inside merges with the outside, and the entity is gone. To persist is to maintain the boundary; to maintain the boundary is to counteract degradation; to counteract degradation is to perform work that biases outcomes toward maintenance. This sequence is not a law of physics; it is a law of identity over time.

This distinction will become crucial when I reach the normative propositions. The "ought" that emerges from the derivation is a constitutive ought, not a moral command. It says: *if* you are a bounded entity in a disruptive world, *then* you must act to maintain your boundary, because ceasing to act is ceasing to be. The imperative is hypothetical in its cosmic form, but categorical for the entity living it. It is not imposed from outside; it is the internal logic of its own existence.

2.4 A Preview of the Eleven Steps

Before we plunge into the detailed exposition, a brief overview of the journey ahead may orient the reader.

Proposition 1 (The Boundary Precondition): To be a determinate entity requires a boundary—a physical or organizational separation from environment. Without this, there is no entity to discuss.

Proposition 2 (The Inevitability of Disruption): In any world that is not perfectly static and homogeneous, interactions across the boundary will degrade its integrity over time.

Proposition 3 (The Primal Distinction): Faced with degradation, systems fall into two classes. Passive conduits are simply carried along by events; active engines harness energy to redirect events.

Proposition 4 (Directed Work): For an active engine, random energy expenditure accelerates disruption. The system's activity must, on net, bias outcomes toward boundary maintenance. Not every act must be efficient, but the total bias must be positive.

Proposition 5 (The Cognitive Minimum): To bias outcomes, the system must discriminate between conditions favorable to persistence and conditions threatening it. This discrimination is a minimal comparison operation—a physical act of sensing a difference.

Proposition 6 (Model-Action Co-Constitution): The comparison constitutes an operational model; the subsequent biasing of outcomes is action. Neither is possible without the other in a persistent system. They arise as a single functional loop.

Proposition 7 (The Constitutive Imperative): The loop is not optional. For the system to continue to exist, it must perform this comparison-action work. The drive to persist—*Hormē*—is not an add-on; it is the condition of being a bounded, far-from-equilibrium entity at all.

Proposition 8 (The Continuous Imperative): Degradation does not pause. The comparison-action loop must therefore be continuous—a *Kybernēsis* process of constant sensing, comparing, and adjusting. Persistence is not a state; it is an ongoing activity.

Proposition 9 (The Life-Agency Isomorphism): Viewed from the inside, the continuous loop is *life*—the self-maintaining activity. Viewed from the outside, it is *agency*—the capacity to redirect causal flow. They are the same physical operation seen through different lenses.

Proposition 10 (The Objectivity of Value): "Good" is constitutively established as whatever fulfills the loop; "bad" is whatever frustrates it. Because the loop is constitutive, value is not a subjective projection but an objective, functional relationship.

Proposition 11 (The Truth-Goodness Isomorphism): To act effectively, the system's model must track reality. Epistemic error and practical failure are the same misalignment. The pursuit of truth and the pursuit of good are a single pursuit.

Each step is a conditional. The antecedent is either one of the initial preconditions or a proposition already established. The consequent is the next theorem. The chain is designed to be transparent: if any step is false, the falsity should be traceable to a specific premise or a specific logical move. The reader is invited to test each link rigorously; this book is organized around that testing.

These eleven propositions are not dogmas. They are the steps of a proof. The remainder of Part II will expand each into a full chapter, with careful justification, exploration of consequences, and a dialog with the strongest objections.

2.5 The Boundary Integral: Formalizing the Necessary Condition

A brief reassurance is in order before proceeding. The reader requires no working knowledge of advanced calculus to follow the argument. The mathematical notation below is simply a formal translation of a highly intuitive concept: that persistence over time is a continuous achievement. Conceptually, one can read the integral as a "historical ledger" of the entity's existence—a running tally where the ongoing work of maintenance acts as deposits, and the ceaseless toll of disruption acts as withdrawals. The math is not here to obscure the philosophy, but to provide it with a precise, quantifiable edge.

The derivation is conceptual, but it yields a formal diagnostic. If persistence requires that the work of maintenance (Φ) exceed the work of degradation (N) over time, then we can write the condition for continued existence as:

$$A(t) = A_0 + \int (\Phi - N) d\tau$$

where $A(t)$ is the organized state of the system at time t, A_0 is the initial boundary capacity, Φ is directed work performed, and N is disruptive work acting upon the system. For the system to remain alive—to persist as a distinct entity—the value of $A(t)$ must remain above a minimum threshold. If the integral becomes too negative, the boundary margin is breached and the entity dissolves.

This equation is not an empirical law like Ohm's Law; it is a formalization of the constitutive necessity. It defines, in operational terms, what it means for a bounded system to persist. It also provides a falsifiability criterion: if a system can be shown to persist indefinitely with Φ consistently less than N, the framework collapses. No such system has ever been documented, and the logic suggests none can be. See Chapter 17: The Mathematics of Boundary Maintenance for the complete formalization of the boundary integral.

2.5.1 The General Law of Identity

This formalization of the living boundary is not an isolated biological equation; it is a specific instantiation of a universal ontological principle: the General Law of Identity.

As established in *The First Cut: How Boundary Logic Derives Physics*, identity is not a static material substance but a historical achievement. The integral $A(t) = A_0 + \int(\Phi - N)d\tau$ represents the literal ledger of existence, summing every moment of repair, energy expenditure, and resistance against decay from the moment of an entity's formation.

In this broader, substrate-neutral context, the variables represent the exhaustive polarity of cosmic dynamics:

- Φ ***(Philia)*** represents the cohesive work that adds to boundary integrity.
- N **(*Neikos*)** represents the dispersive tendency or decay that subtracts from it.

The requirement that cohesive work must offset disruptive flux to maintain a threshold of integrity applies to *any* persisting pattern in a non-inert world. The differential shadow of this integral—the instantaneous rate of maintenance $dA/dt = \Phi - N$—is the universal mathematical structure of persistence. It is the exact form that governs diverse physical realities:

- In thermodynamics, it describes the entropy balance.
- In population dynamics, it tracks births minus deaths.
- In economics, it dictates investment against depreciation.

To exist as a determinate entity across time—whether a cell, a star, a corporate organization, or a cognitive model—is to satisfy this fundamental mathematical condition. While the present derivation focuses specifically on the threshold of life and minimal agency, the universal scope of this boundary logic, and its formal derivation of physics and thermodynamics, is explored fully in The First Cut.[2]

2.6 The Role of Empirical Science: Intersection, Not Foundation

A priori biology does not ignore empirical science; it intersects with it at every point. The derivation predicts that any living system will exhibit a comparison-action loop, a constitutive imperative, and a structural coupling of truth and goodness. These predictions can be tested against the observed world. Where empirical science has studied the mechanics of persistence—in Schrödinger's *What Is Life?*, in Maturana and Varela's autopoiesis, in Friston's active inference—it has discovered precisely the architecture that the deduction demands.

But the deduction does not depend on these empirical findings for its validity. It stands on its own logical feet. The empirical findings corroborate; they do not establish. This is the reverse of the usual scientific procedure, where theory is built from data. Here, the theory is derived from first principles, and data provide confirmation. A mismatch between deduction and observation would refute the deduction. So far, the match is exact.

2.7 How to Read This Book: A Roadmap

This book is written to be read in order. The derivation is cumulative; Proposition 7 cannot be understood without Propositions 1–6, and the objections later in the book will make little sense without the arguments they challenge.

Part II presents the derivation itself, one chapter per proposition. Each chapter is structured identically: a clear statement of the proposition, a careful unpacking of its logical and physical content, an exploration of its immediate implications, and a concluding section briefly clarifying objections.

[2] See Eli Adam Deutscher, *The First Cut: How Boundary Logic Derives Physics*, Neo-Pre-Platonic Press, 2026, `https://neopreplatonic.com/papers/First_Cut/`.

Part III deepens the mechanical picture. I examine the *Kybernēsis* in detail, distinguish teleonomy from teleology, and show that Darwinian evolution emerges deductively from the logic of finite lifespans and replication.

Part IV formalizes the derivation and draws out its predictive power: the boundary integral equation, the resolution of asymmetric risk in evolutionary psychology, the ontological mandate of minimal determinism, and the dynamics of nested *Hormic* architectures.

Part V maps the derivation onto the history of science. I will show that major figures—Schrödinger, Gánti, Maturana and Varela, Friston, Jonas—captured critical pieces of the architecture, and that their limitations become visible precisely at the points where they lacked the deductive method.

Part VI is the book's stress test. Objections that received summary treatment in Part II are here developed into full dialogues. The Is-Ought gap, the challenge of AI, the paradox of suicidal agents, the Moorean Open Question—each receives sustained, patient rebuttal.

Part VII connects the derivation to the wider system of Neo-Pre-Platonic Naturalism, showing how the logic of the boundary undergirds an entire philosophy of mind, ethics, and navigation.

Above all, this book is an invitation to think alongside the author. The argument is laid out in full view, with all its premises exposed and all its joints visible. The reader is not asked to accept anything on faith, only to follow the chain and test each link. If the chain holds, the conclusions are inescapable. If it breaks, the reader will have found the exact point of breakage, and that too is a gain for clarity.

With the method now clarified and the map in hand, we turn to the first step: the boundary itself.

Part II

The Formal Derivation

Chapter 3

The Boundary Precondition

Every thing is a *this* against a *not-this*. A rock, a cell, a nation, a thought—whatever possesses determinate identity does so by virtue of being separated, in some manner, from everything it is not. This separation is the boundary, and it is not a biological luxury. It is the minimal logical condition for anything to exist as a distinct entity at all. The first proposition of the derivation is therefore not a discovery about living systems but a recognition of the prerequisite of determinacy itself.

Proposition 1 (The Boundary Precondition). To possess determinate identity, an entity must be distinct from its environment. Therefore, to "be" something requires a physical or organizational boundary that separates it from everything it is not.

This chapter unpacks that proposition. We begin by clarifying what a boundary is in this minimal sense, then show why it is logically inescapable. If the proposition holds, it is the foundation on which the remaining ten steps will be built. If it crumbles, the entire derivation collapses with it.

3.1 The Logic of Distinction

Consider the statement "X exists." What must be true for that statement to have meaning? At minimum, X must be discriminable from what is not X. If there is no difference between X and its complement—if the universe were a perfectly featureless homogeneity, a "Parmenidean One"[1]—then the word "X" would pick out nothing. There would be no X to refer to.

[1]See Parmenides, "Fragment 8," in *Die Fragmente Der Vorsokratiker*, 6th ed., ed. Hermann Diels and Walther Kranz (Weidmann, 1951), fragment 8 (DK 28 B8).

The principle is as old as Parmenides and as contemporary as information theory: to be is to be distinct.[2]

This is not an empirical claim. It is not that we have looked around the universe and noticed that everything seems to have boundaries. It is that the very concept of an *entity*—a thing with identity conditions—presupposes a principle of demarcation. A thing is a thing only if there is a fact of the matter about where it ends and the rest of the world begins. Without that fact, we are not talking about a thing; we are talking about an undifferentiated background.

The boundary, then, is the ontological realization of the logical law of non-contradiction. To say that a rock is a rock is to say that it is not a tree, not air, not water, not the ground beneath it. The "not" must be anchored somewhere. The surface of the rock—where its mineral structure gives way to atmosphere—is that anchor. It is the physical trace of the logical distinction between *this* and *not-this*.

One might object that this is mere semantics. "We define X by contrast, but that's just how language works; it doesn't mean X has a boundary in the world." The reply is that language works that way because reality works that way. If reality were not carved at the joints—if there were no genuine discontinuities, no differences that exist independently of our descriptions—then language could not carve it. The fact that we can successfully refer to individual rocks, organisms, and planets is evidence that the world contains genuine boundaries. The logical requirement and the ontological fact coincide.

3.2 What Counts as a Boundary

A boundary, in the sense used here, is any physical or organizational barrier that separates an inside from an outside. It need not be rigid. It need not be impermeable. It need not be visible to the naked eye. It need only constitute a difference that matters for the entity's continued existence as itself.

Physical boundaries are the easiest to visualize. The membrane of a cell, the skin of an animal, the surface of a rock—these are spatial discontinuities where the material composition or density changes sharply. They are

[2]Norbert Wiener, *Cybernetics: Or Control and Communication in the Animal and the Machine* (MIT Press, 1948).

the boundaries of everyday objects. They function as boundaries because interactions that cross them change the entity in ways that interactions within them do not.

Organizational boundaries are subtler but equally real. A corporation has no physical skin; its boundary is a legal and functional one, defined by contracts, decision-making structures, and the flow of money. A language community has a boundary defined by mutual intelligibility and shared norms. An ecosystem's boundary is the limit within which energy and nutrient cycles are largely self-contained. In each case, the boundary is not a surface but a threshold in a relational network—a condition that distinguishes insiders from outsiders in a way that is causally consequential.

The present derivation is substrate-neutral. It insists only that *some* boundary exists that constitutes the entity as a distinct player in its domain. The precise form of the boundary will depend on the kind of entity under study, and later propositions will specify what kind of boundary is required for life. But at this first step, the requirement is minimal: no boundary, no entity.

3.3 The Transcendental Argument: Why the Boundary Cannot Be Denied

The strongest way to establish a first principle is to show that denying it is self-undermining. Suppose someone asserts: "Boundaries are not necessary for existence. There exist things that are wholly continuous with their environment, with no separation whatsoever." What would it mean for such a thing to exist? If there is no separation, then there is no fact about which part of the world is the thing and which part is the environment. The supposed entity is identical to its surround. But then it is not an entity at all; it is merely a region arbitrarily designated by the speaker. The assertion reduces to: "I can point to a patch of homogeneous fluid and call it a thing." That gesture does not demonstrate the existence of a boundaryless entity; it demonstrates the speaker's capacity to draw a boundary in thought. The very act of pointing *imposes* a boundary, presupposing the principle it claims to refute.

The argument is transcendental in the Kantian sense.[3] It asks for the conditions of possibility of any talk about entities, and finds the boundary among them. Any attempt to deny the boundary must employ concepts that already presuppose boundaries—the concept of the denying speaker, the concept of the proposition being denied, the concept of the physical world in which the denial occurs. The would-be skeptic stands on a raft made of boundaries while insisting that boundaries do not exist.

This does not mean that all boundaries are crisp or eternal. The coastline is famously fractal; the edge of a cloud is a matter of convention. But the fact that a boundary is fuzzy does not mean it is absent. It means the boundary is a graded transition rather than a step function. That transition is still a difference, and that difference is what makes the cloud a cloud and not the sky. Vagueness is not the same as non-existence.

3.4 The Scope: Physical Entities in Time

One clarification of scope is necessary. This book concerns life, and life is a physical phenomenon. The entities we are ultimately interested in are physical systems that exist in time and are subject to causal interaction. The boundary precondition, as we will use it, applies paradigmatically to such systems. Abstract entities—numbers, sets, logical truths—are not physical, and their mode of existence does not involve causal interaction or degradation. They are not the subject of this derivation. If they have boundaries (as the set of primes is distinct from the set of composites), those boundaries are conceptual, not physical. Nothing in the argument depends on extending the analysis to abstracta; the derivation is exclusively about concrete, spatiotemporally located systems.

With that scope in place, we proceed to the clarification.

3.5 Clarifications and Forward References

A few predictable concerns arise immediately from the boundary precondition. We address them here only in the briefest form, to remove obstacles that might prevent the derivation from proceeding. The full steelmanned versions of these objections—together with several others—are

[3]Immanuel Kant, *Critique of Pure Reason*, trans. Paul Guyer and Allen W. Wood (Cambridge University Press, 1998).

developed and answered in detail in Part VI (see Chapter 26: The First Wall: Objections to the Boundary).

Circularity. The boundary and the identity it bounds are co-constitutive primitives, not a vicious circle. Just as geometry takes "point" and "line" as mutually defining axioms, this derivation takes the identity-boundary pair as its irreducible starting point.

Fields and fuzziness. A boundary need not be a sharp wall; it can be a graded transition, a statistical threshold, or a source-defined region. The only requirement is that there is a fact of the matter about which entity we are discussing, and therefore some causal contrast with its surround.

Elasticity of the concept. The boundary is deliberately defined functionally—it is whatever separation constitutes an inside/outside difference that matters for persistence. This substrate-neutrality is a strength, not a vacuity, allowing the derivation to apply across all possible material realizations.

Observer-dependence. That some boundaries are conventional (clouds, coastlines) does not entail that all are. The observer who draws the convention is himself a bounded organism; real boundaries are a precondition for the very act of drawing conventional ones.

Holobionts and symbiosis. The apparent blurring of organismal boundaries reveals not the absence of boundaries but their *nesting*. A holobiont is a higher-order organizational boundary composed of many lower-order boundaries. This modular, layered architecture is a central feature of life, not a counterexample to the boundary precondition.

Each of these points receives a rigorous, no-retreat defense later in the book. For now, they are flagged only so that the derivation may continue without unnecessary friction.

3.6 The Boundary as the Foundation

What has been established? That to exist as a determinate thing is to be bounded. This is not an empirical guess; it is the condition on which any talk of entities is possible. The boundary may be physical or organizational, sharp or graded, simple or nested. But it must be there. Without it, there is no entity to talk about, no agent to act, and no problem of persistence to solve.

The boundary is the minimal ontological unit. It is not life, but it is the necessary precondition for life. The next step is to observe that boundaries, once established, do not float in a vacuum. They are embedded in an environment, and that environment is not inert. Interactions happen. And interactions degrade.

Chapter 4

The Inevitability of Disruption

A boundary is not a fortress. It is an interface, and to be an interface is to be touched. In any world that is not inert, those entities will interact, and their interactions will cross whatever boundaries they possess. The crossing leaves a mark. Over time, marks accumulate, and the boundary that once cleanly separated inside from outside begins to blur, erode, or break. Disruption is not an occasional misfortune; it is the permanent condition of being a bounded thing in a populated universe.

Proposition 2 (The Inevitability of Disruption). Bounded entities are subject to continuous interaction. These interactions degrade the integrity of the boundary over time.

This chapter defends that proposition. We examine why interaction is inescapable, why degradation is its structural consequence, and why the point holds regardless of the material composition of the boundary or the specific physics of the world.

4.1 The Inescapability of Interaction

To be bounded is not to be isolated. A boundary is defined by its environment as much as by its interior; the two are complementary. But if the environment exists, then by definition it is not identical to the interior. There is a difference, and across that difference, something can happen.

What must that "something" be? If the universe is not perfectly frozen, perturbations will propagate and encounter the boundary. A "perfect reflector" is a phantom concept; to exist in a world is to be causally connected to it. Every interaction is a displacement or exchange of energy that alters the state of the boundary, even if only infinitesimally. Because the boundary is a physical structure rather than a mathematical abstraction, it is inherently deformable. Interaction is the tax reality levies on existence, and it is paid in the currency of disruption.

To exist as a physical entity is to be causally connected to the rest of the universe. An entity does not float in a sterile void; it is submerged in a continuous bath of ambient dynamics. Even in the deepest interstellar vacuum, a boundary is bombarded by the cosmic microwave background. In any populated environment, it is struck by photons, jostled by atmospheric molecules, and subjected to gravitational shear and thermal fluctuations. Because the entity is causally connected to the world, these interactions transfer energy across its boundary. And because the boundary is a specific, highly organized structural arrangement, this continuous, undirected transfer of energy will inevitably nudge its material out of its initial state. The disruption is not a discrete accident; it is the continuous tax levied by physical existence.

The point is not that disruption always happens on a particular timescale, or with a particular intensity. It is that in any world with multiple distinct entities and some dynamics, disruption is possible. And over time, what is possible and undirected becomes statistically certain. The boundary is subject to a kind of random bombardment, and random bombardment does not, on net, improve order.

4.2 The Logic of Degradation

That disruption is inevitable does not yet establish that it is degrading. Could not some interactions fortify the boundary, leaving it stronger than before? In principle, yes—if the environment were directed. If every interaction were perfectly aimed at restoring the boundary's specific configuration, then disruption would never become degradation. But the environment, by our third starting assumption, is blind. It possesses no goal, no aim, no preference for the maintenance of any particular arrangement.

Given that the environment is undirected, the question reduces to a statistical one. A boundary is a specific, organized configuration. There are vastly more ways for that configuration to be altered than for it to be preserved or improved. A random perturbation is overwhelmingly likely to move the boundary material away from its organized state, simply because the volume of phase space corresponding to "organized boundary" is minuscule compared to the volume corresponding to "disorganized non-boundary."

To make this concrete, consider a sandcastle that is exactly one grain of sand away from structural perfection. A gust of wind sweeps across the beach. Is it physically possible that the wind carries exactly one grain of sand, traveling at precisely the right velocity, and deposits it flawlessly into the missing slot without disturbing a single other grain? Yes. It violates no laws of physics. But there is exactly one trajectory and velocity that achieves this outcome. In contrast, there are trillions of wind trajectories that will strip grains away, pit the walls, or level the tower entirely. The degradation of the castle is not caused by a malicious, goal-oriented wind; it is driven by the sheer, crushing mathematical asymmetry between the single way to build and the near-infinite ways to destroy. This is the statistical proof of disruption.

Thus degradation is not a separate force that needs to be posited. It is the expected consequence of undirected interaction with a complex structure. If the universe were so finely tuned that every random interaction happened to exactly maintain every boundary, that would be a cosmic conspiracy of staggering improbability. In any world without such a conspiracy—that is, in any world recognizable as real—boundaries degrade.

The Second Law of Thermodynamics is not an axiom of this derivation; rather, it is merely the local, empirical shadow of this universal statistical logic.[1] By appealing only to the logic of organized structures under random perturbation, the argument frees itself from concepts of entropy or free energy. If a reader inhabited a universe with no concept of thermodynamic entropy but still experienced the fact that untouched houses fall into disrepair and that constant effort is required to keep a garden from reverting to weeds, that reader could follow the argument. The statistical logic is sufficient.

[1]This echoes Boltzmann's statistical interpretation of entropy, wherein the Second Law is not a fundamental mechanical force, but a statistical certainty driven by the overwhelming asymmetry between the number of microstates corresponding to disordered versus ordered macrostates. See Ludwig Boltzmann, "Über Die Beziehung Zwischen Dem Zweiten Hauptsatze Der Mechanischen Wärmetheorie Und Der Wahrscheinlichkeitsrechnung Respektive Den Sätzen Über Das Wärmegleichgewicht," *Sitzungsberichte Der Kaiserlichen Akademie Der Wissenschaften. Mathematisch-Naturwissenschaftliche Classe* 76 (1877): 373–435.

4.3 Sources of Disruption

Disruption comes from both outside and inside. External disruption is the most obvious: collisions with other entities, radiation from distant stars, chemical attack from corrosive substances in the environment, temperature fluctuations that cause thermal expansion and contraction. The boundary is a front line, and the front line is always under fire.

Internal disruption is subtler but equally universal. Inside any bounded entity, processes occur. If the entity is not entirely homogeneous and static, then internal dynamics will produce byproducts—wastes, structural defects, imbalanced distributions—that accumulate. Even a perfectly sealed box containing a simple chemical system will eventually reach equilibrium, but during the approach to equilibrium, local gradients can damage internal structural components. In a complex, far-from-equilibrium system, internal processes are ceaseless, and they produce internal wear.

The distinction between internal and external disruption will become critical later in the derivation, when we examine how living systems must manage both. For now, the point is simply that disruption is not a one-source problem. Even an entity that perfectly sealed its boundary against all external forces would still face the slow accumulation of internal disorder.

4.4 Timescale and the Fallacy of the Static

One might object that some entities seem to persist unchanged for immense spans of time. A crystal of quartz buried in granite can last for eons. A proton, so far as we know, is stable across timescales longer than the age of the universe. Does not such persistence refute the claim that disruption is inevitable?

The reply is that timescale does not refute inevitability; it only quantifies it. Proposition 2 does not claim that disruption is *rapid*. It claims that over sufficient time, interactions will degrade the boundary, and that no finite entity possesses an infinite resistance to every possible interaction. The quartz crystal inside the granite is protected from many external forces, but it is still subject to the slow bombardment of cosmic rays, the gradual diffusion of trace elements, and eventually—when the granite weathers—the full force of erosion. The proton may be stable, but protons are not bounded systems with an inside and an outside in the organizational sense; they are elementary particles, and the derivation's notion of boundary does not apply in the same way. The entities we care about—organisms, cells, ecosystems—are composite structures with internal organization, and that organization is statistically fragile.

More fundamentally, the objection mistakes a local, temporary stasis for a universal condition. A crystal in a saturated solution may indeed maintain its boundary indefinitely *if the environment remains perfectly saturated.* But the environment is itself a dynamic system; perfect, unchanging saturation is an idealization that reality only approximates. Over deep time, the environment shifts. The crystal is in a metastable state, not an absolute equilibrium with all possible perturbations. Its apparent permanence is a timescale artifact.

4.5 Clarifications and Forward References

The inevitability of disruption invites several natural challenges. Some entities appear to persist without apparent effort. Some systems seem closed enough to escape interaction. And some boundaries might be imagined that are perfectly self-repairing. We address these concerns briefly here, reserving their full steel-manned treatment for Part VI (see Chapter 26: The First Wall: Objections to the Boundary).

Crystals in saturated solution. A crystal in equilibrium with its solution persists only while the environment is held artificially constant—by an experimenter whose own *Hormē* maintains the setup. In an unmanaged, non-inert world, that stability is temporary.

The perfectly closed system. A box with perfectly reflective walls and a static interior is either the entire universe or an idealization that ignores internal differentiation and the external forces that must maintain the box itself. In any world with multiple distinct entities and some dynamics—the only kind of world where life can arise—truly closed systems are idealizations, not realities.

Self-repairing boundaries. A boundary that spontaneously reseals, such as a lipid bilayer, is already performing directed work of a sort, anticipating the comparison-action loop of later propositions. The question of whether such self-repair can occur without any metabolic support is deferred; what matters now is that *absent* such active maintenance, boundaries degrade.

The objection from contingent physics. The charge that degradation relies on the Second Law of Thermodynamics mistakes a special case for the general principle. Disruption follows from the mathematics of blind flux applied to a specific arrangement, and from the simple fact that to exist in a world is to be causally connected to it.

4.6 The Foundation Laid, the Fork Ahead

We now have two truths in hand. A bounded entity exists as a distinct thing (Proposition 1). That boundary is subject to continuous disruption from both external and internal sources (Proposition 2). The conjunction of these two propositions creates the existential problem that the rest of the derivation must solve: how can a bounded entity persist over time when its boundary is continuously being degraded?

The answer cannot be "it just does." Passive endurance is a temporary reprieve, not a strategy. If an entity is to persist beyond the brief window during which its initial highly organized configuration survives random bombardment, it must do something. The something it must do will be the subject of Propositions 3 and 4.

The next chapter draws the primal distinction: between those systems that are simply carried along by the causal flow—passive conduits—and those that actively resist it—active engines. The distinction is the first fork in the ontological road, and from it, agency will emerge.

Chapter 5

The Primal Distinction

We have established that any bounded entity faces continuous degradation of its boundary (Proposition 2). That degradation is not a special law; it is the expected consequence of existing in a world with other things. The question now is: what can a bounded entity do about it?

The answer divides the universe of bounded things into two and only two kinds. Either the entity does nothing—it is simply carried along by events, its boundary persisting only so long as the environment happens not to destroy it. Or the entity does something—it intervenes in the stream of events, expending its own stored capacity to deflect that stream. The two kinds are categorically distinct. There is no middle ground.

Proposition 3 (The Primal Distinction). The requirement of boundary maintenance creates a primal distinction in physical systems. **Passive conduits** are inert aggregates; they are carried along by the causal flow and are unable to deviate from its course. **Active engines** are dynamic systems that harness external energy or expend internal energy to redirect the causal flow.

This chapter draws that distinction clearly, defends it against the charge of false dichotomy, and prepares the ground for the detailed mechanics of active engines that will occupy the next several chapters.

5.1 The Concept of Causal Flow

Before we can talk about redirecting "causal flow," we must clarify what the term means. Causal flow is simply the ordered operation of physical law—the way events unfold from prior events according to whatever regularities the world possesses. In a deterministic world, causal flow follows a law-bound trajectory; given the state of the world at one moment, the next moment follows uniquely. In a probabilistic world, causal flow is a distribution; given the state of the world, the next moment is drawn from a probability distribution that is itself lawfully constrained. In every world that is not pure chaos, there are such lawful constraints.

The term "causal flow" does not commit the argument to any particular metaphysics. It does not require hard determinism, nor does it require a specific physics. It requires only that the world has some degree of regularity—that the same conditions tend to produce the same outcomes. Without that, no entity could rely on anything, and the very concept of "persistence" would lose its meaning, because what happens next would be utterly unrelated to what came before. We will return to this requirement later in the chapter, for it carries a profound implication about the universe that life inhabits.

For now, think of causal flow as the current of events. Everything that happens is part of it. The question Proposition 3 asks is: does a given bounded entity merely *float* on this current, or does it *swim*?

5.2 Passive Conduits

A passive conduit is a bounded entity that does nothing to alter the causal flow in which it is embedded. It is acted upon, but it does not act. If a rock rests on a hillside, gravity pulls it downhill, rain wears its surface, and wind chips at its edges. The rock's boundary—its mineral surface—is continuously degraded. The rock does not resist. It does not expend any energy of its own to shore up its structure, to flee the rain, or to anchor itself against gravity. It is a piece of the causal flow, indistinguishable in its passivity from any other inert aggregate.

This is not to say that passive conduits cannot persist for long periods. A rock in a desert cave may survive for millions of years. But this persistence is entirely a matter of good fortune—an environment that happens, for a time, to exert minimal disruptive force. The rock is not *maintaining* its boundary. It is simply benefiting from the absence of hostile conditions. When the conditions change, the rock will yield. It has no capacity to do otherwise.

The key feature of a passive conduit is that its internal organization contributes nothing to its own trajectory. Any energy within it—thermal vibrations, chemical potential—is not harnessed to alter its state. If a random vibration happens to seal a microcrack, that is an accident of the same statistical kind that might equally have widened the crack. There is no bias in the outcomes, because there is no mechanism that links internal states to causal effects. The entity is, in a precise sense, blind to its own fate.

5.3 Active Engines

An active engine is a bounded entity that does not merely undergo the causal flow but intervenes in it. It draws upon a store of internal energy—or captures energy from its environment—and deploys that energy to alter the course of events.

The term "engine" is deliberate. An engine is a device that converts energy into directed work. The active engine is a natural engine: it burns fuel—chemical, photonic, thermal—and uses the released energy to push, pull, build, or repair. It is not a passive victim of its surroundings; it is an active participant in the unfolding of events.

Consider a simple case: a bacterium spinning its flagella. The bacterium is not being pushed by an external mechanical force. It is reorganizing its internal energy to alter its position. The causal flow without the bacterium would have seen the surrounding fluid move only according to external pressure and diffusion; the causal flow with the bacterium is different, because the bacterium has injected its own causal power into the stream. It has redirected events.

This capacity to redirect causal flow is the signature of an active engine. It does not require consciousness. It does not require a brain. It requires only that the entity's internal organization be such that its energy expenditure alters the default physical outcome. The point for now is that active engines do not merely occur; they *act.*

5.4 The Exhaustiveness of the Distinction

The distinction between passive conduits and active engines is logically exhaustive. Every bounded entity in an interacting world either redirects causal flow or it does not. There is no third option. An entity cannot "half" redirect. It either possesses a mechanism that deviates from the default trajectory, or its outcomes are entirely at the mercy of the undirected causal flow.

Even a pathological system—an entity that redirects causal flow to accelerate its own destruction—would still be an active engine. The division is as clean as "moves" vs. "does not move" in the simplest physical sense. Either the entity's internal dynamics are causally relevant to its trajectory, or they are not. There is no conceptual space between.

5.5 The Primacy of the Distinction

Why is this distinction "primal"? Because it is the first branching point after the bare preconditions of existence. Proposition 1 gave us the boundary; Proposition 2 gave us its vulnerability. Now, with Proposition 3, the ontological road forks. Down one path lies everything that is merely matter, passively unfolding according to laws it does not harness. Down the other path lies everything that has the capacity for agency.

Notice that the distinction does not yet mention metabolism, reproduction, evolution, or any of the usual biological criteria. It is more fundamental than any of those. An entity could be an active engine without reproducing, without evolving, without being made of cells. It need only redirect causal flow.

5.6 The Ontological Floor: Minimal Determinism

Before proceeding, we must note a profound implication that the distinction carries with it. For an active engine to be possible, the causal flow must be redirectable. And to be redirectable, it must have a degree of predictability. In a universe of pure chaos—where no regularities hold, where the same action can produce arbitrarily different outcomes—no entity can bias events toward any outcome. The concept of "harnessing energy to redirect causal flow" is meaningless if the flow has no consistent direction.

Therefore, the mere existence of an active engine—any active engine, anywhere—constitutes proof that its universe possesses what we may call *minimal determinism.* The universe need not be rigidly clockwork. It may contain genuine indeterminacy, quantum or otherwise. But it must contain enough lawful regularity that some actions have reliable consequences, so that an entity can learn—through evolution or through individual adaptation—that doing X produces a different outcome than doing Y.

This insight is not a side note. It will return later in the book as a major ontological result: life is not a violation of physical law but the ultimate exploitation of physical law. The very possibility of an active engine demonstrates that the world is structured, that its Logos—its inherent order—is, at least in part, discoverable and usable. A universe without minimal determinism is a universe in which boundaries dissolve at random, and no engine can effect outcomes. Since we observe active engines—ourselves among them—we know we do not inhabit such a universe.

5.7 Clarifications and Forward References

The primal distinction is simple, but its simplicity invites questions. Several natural candidates appear to sit between passive conduits and active engines, and we pause here only long enough to clarify why they do not blur the line. The full steel-manned treatment of these cases, together with others, resides in Part VI (see Chapter 27: The Second Wall: Objections to the *Kybernēsis* Core).

Hurricanes and whirlpools. A hurricane maintains a coherent structure for days, but it does not *redirect* the causal flow; it *is* the causal flow—the visible shape of a pressure gradient equalizing itself. There is no internal mechanism that alters the system's dynamics relative to the gradient. When the gradient dissipates, the structure vanishes without resistance. The same holds for whirlpools and other dissipative structures: they are the transient trace of energy flowing downhill, not engines that push back.

Machines. A solar-powered water pump is indeed an active engine by the definition of Proposition 3, and the derivation embraces this classification. The Primal Distinction is the *first* cut, not the final definition of life. Later propositions—specifically Proposition 7 on intrinsic versus extrinsic *Hormē*—will distinguish between engines that maintain their own boundaries and engines that serve an external goal. The pump belongs in the active category; the refinement comes later in the derivation.

Rivers. A river carves its channel and appears to "preserve" its path, but it has no boundary to maintain and no mechanism that alters its flow. It simply follows gravity, and the erosion of a channel is a byproduct of that downhill motion, not an intervention by a bounded system. The river is part of the causal flow, not an agent redirecting it.

Flames. A flame draws in fuel and sustains a reaction zone, but it does not modulate its burn rate to alter outcomes; it accelerates the consumption of its fuel until it exhausts the supply and dies. There is no modulation, no mechanism that dampens combustion when fuel is scarce. It is a runaway reaction, not a regulated engine, and therefore remains on the passive side of the distinction.

5.8 The Fork in the Road

We arrived at Proposition 3 with a bounded entity facing continuous degradation. We now see that the entity has two and only two possible modes of existence. It can be a passive conduit, allowing the causal flow to carry it wherever it will, persisting only while circumstances happen to be benign. Or it can be an active engine, expending energy to push back, to deflect the flow, to carve a channel through a stream that would otherwise sweep it away.

The passive conduit is the default state of mere matter. The active engine is the first glimmer of something different—the capacity to intervene in reality. It is not yet life; we have not yet defined the specific kind of active engine that constitutes a living being. But it is the category within which life will be found.

But what must an active engine do to *persist*? It is not enough merely to expend energy. Random expenditure would accelerate degradation, like a man thrashing in quicksand. For the engine to survive, its expenditure must have a specific orientation. It must be *directed work*. That is Proposition 4, and to it we now turn.

Chapter 6

Directed Work

An active engine does not merely drift. It expends energy. But energy expenditure alone guarantees nothing. A frantic discharge of effort that does not discriminate between building the boundary and tearing it down will, on average, accelerate the engine's dissolution. To persist, the engine must aim—not in the sense of conscious intention, but in the operational sense that its actions must, on balance, push the causal flow in a direction that favors the boundary's continuance. This is the requirement of directed work.

Proposition 4 (Directed Work). Random energy expenditure accelerates disorder. To effectively maintain a boundary, the system's energy expenditure must, on net, bias causal outcomes toward its own continued persistence. Not every action need be optimal; some may be neutral or even counterproductive, provided the overall effect sustains the boundary above the threshold of dissolution.

This chapter unpacks the concept of directed work. We examine why randomness fails, what it means for work to be directed without invoking conscious purpose, and how the requirement of a net bias navigates the space between rigid determinism and chaotic expenditure.

6.1 The Failure of Random Expenditure

Suppose an active engine faces a degrading boundary. It possesses a store of internal energy—chemical bonds, ion gradients, mechanical tension—and it releases that energy into its environment. If the release is entirely uncorrelated with the boundary's condition, what happens?

The environment is indifferent to the engine's persistence. The engine's boundary is a specific, highly organized configuration. Random pushes from the engine's own activity are, from the boundary's perspective, indistinguishable from random pushes from the environment: they are perturbations that are overwhelmingly more likely to disrupt the organized structure than to fortify it. The statistical logic that guarantees degradation from external interactions (Proposition 2) applies equally to internal actions if those actions are not functionally coupled to the boundary's state. Random expenditure is just another source of noise, and noise erodes.

6.2 The Concept of Directed Work

"Directed work" is work whose effect is, on net, to maintain or restore the boundary. The emphasis is on *net*. The derivation does not require that every individual action be beneficial. An active engine may stumble, make errors, or perform actions that are neutral in their effect on persistence. It may even perform actions that are temporarily harmful—the bacterium that swims briefly down a gradient before correcting course, the repair enzyme that occasionally misbinds. What is required is that over the relevant timeframe, the total vector of the engine's activity points toward boundary maintenance. The bias must be positive, not perfect.

This notion of net bias is crucial for the derivation's biological plausibility. Real organisms are not flawless optimizers. They carry vestigial structures, they exhibit spandrels, they sometimes make costly errors. A framework that demanded perfection would be falsified by the first observed mistake. A framework that demands only a net positive contribution to persistence can accommodate messiness while still drawing a sharp line between the active and the inert.

The "direction" of the work is not a mystical quality. It is a statistical correlation between the engine's energy deployment and the boundary's continued integrity. If, after some interval, the engine's expenditures have repaired more damage than they have caused, the work was directed. If they have caused more damage, the work was misdirected—and the engine will, by the logic of Proposition 2, eventually cease to exist. Persistence over time is itself the proof that the work is, on net, directed. A system that endures is a system whose activity is functionally adequate.

6.3 Teleonomy Without Teleology

The phrase "directed work" risks the charge of teleology—the idea that outcomes are guided by future purposes. This charge must be met directly. I posit no future-pulling force, no Aristotelian final cause that reaches backward in time to shape present action. The direction of the work is entirely a consequence of the system's present organization and its causal coupling to the environment.

The appropriate term is *teleonomy.* Teleonomy describes apparent purpose that arises from feedback-stabilized mechanisms.[1] A thermostat does not "seek" a temperature in the sense of being pulled by a future state; it simply has a physical structure such that when the temperature drops, a bimetallic strip bends, closing a circuit, turning on a heater. The "goal" is a property of the mechanism's organization, not a mysterious force. Similarly, an active engine that biases outcomes toward persistence does so because its internal architecture—evolved, designed, or self-organized—links specific environmental signals to specific energy deployments that have, in the past, conduced to survival. The correlation between action and persistence is historically grounded, not metaphysically guaranteed.

This distinction will deepen when we reach the cognitive minimum (Proposition 5) and the constitutive imperative (Proposition 7). For now, it suffices to say that "directed" is a description of the functional relationship between energy expenditure and boundary maintenance, not an invocation of conscious intention or cosmic purpose.

6.4 The Threshold of Dissolution

The requirement of net bias is not infinitely elastic. A boundary has a certain tolerance—a margin of error beyond which accumulated damage becomes irreversible and the entity ceases to exist. The directed work must keep the system above this threshold. Formally, we can say that the integral of boundary-maintaining work must continuously exceed the integral of boundary-disrupting forces; but even without formalizing, the concept is intuitive. A boat with a slow leak must be bailed at a rate that outpaces the leak, on average. Occasional pauses for rest are fine if the boat is not swamped in the meantime.

[1] Colin S. Pittendrigh, "Adaptation, Natural Selection, and Behavior," in *Behavior and Evolution*, ed. Anne Roe and George Gaylord Simpson (Yale University Press, 1958), p. 394

The threshold gives teeth to the concept of directed work. It is not enough that the engine sometimes, in good moments, does something helpful. The help must be sufficient to offset the damage. A system that biases outcomes toward persistence but with too weak a bias will still dissolve. The bias must cross a minimal effectiveness threshold, and that threshold is set by the harshness of the environment. In a gentle environment, a sloppy bias may suffice; in a harsh one, only a precise and powerful bias will do. The variation explains why organisms in extreme environments often display exceptionally tight regulatory control, while those in stable, resource-rich environments can afford more slop.

We will formalize this threshold later in the book with the boundary integral equation. For now, the qualitative point stands: directed work is work that keeps the boundary above water.

6.5 Clarifications and Forward References

Directed work is the first proposition in the derivation that carries an empirical edge, and it invites several natural questions about teleology, measurement, and the prevalence of apparently non-adaptive traits. A brief clarification here removes immediate obstacles; the full steel-manned treatment of these challenges belongs to Part VI (see Chapter 27: The Second Wall: Objections to the *Kybernēsis* Core).

The charge of teleology. "Directed" does not invoke final causation. The derivation uses the term in the same functional shorthand as a thermostat "directed" toward a set-point: the full causal story is available in proximate physical terms. A bacterium's chemoreceptor does not aim at a future state; it physically couples to the environment in a way shaped by past selection. The distinction between teleology and teleonomy is developed in depth in Chapter 15: Teleonomy Without Teleology and defended against the strongest objections later.

The vagueness of the threshold. The threshold of dissolution is quantifiable and will be formalized through the boundary integral equation in Part IV. Empirically, it corresponds to well-known collapse points—membrane potential, blood loss volume, minimum viable population size—that are already measured in the life sciences. The concept is not vague; it is precise enough to be falsifiable even before the formalization arrives.

The spandrel problem. The derivation does not claim that every trait must be optimally directed. It claims only that the organism's *total* activity must, on net, bias outcomes toward persistence. Spandrels, vestigial organs, and junk DNA are the tolerated freight of a system whose core directed-work machinery generates enough surplus to subsidize them. The existence of neutral traits confirms the derivation's notion of a boundary margin, not a demand for architectural perfection.[2]

Random actions that happen to work. A system whose boundary-maintaining outcomes are purely fortuitous is, statistically, a passive conduit, not an active engine. Over time, random twitches produce vastly more harmful results than helpful ones; the asymmetry of blind flux against specificity guarantees a net negative drift. Persistent survival is itself the evidence of systematic coupling, not a lucky run.

6.6 The Engine Begins to Steer

With Proposition 4, the active engine acquires its first definitive characteristic: it does not merely expend energy; it expends energy in a way that is systematically tethered to its own persistence. This tethering is what gives the engine its causal signature—the ability to tilt the playing field in its own favor, however slightly, against a universe that tilts it back.

But how does the engine know which way to tilt? How does it discriminate between conditions favorable to persistence and conditions hostile to it? The mere requirement of net bias cannot be met without a mechanism that registers the difference. That mechanism is the subject of Proposition 5: the cognitive minimum.

[2]Stephen Jay Gould and Richard C. Lewontin, "The Spandrels of San Marco and the Panglossian Paradigm: A Critique of the Adaptationist Programme," *Proceedings of the Royal Society of London B* 205, no. 1161 (1979): 581–98, `https://doi.org/10.1098/rspb.1979.0086`.

Chapter 7

The Cognitive Minimum

An active engine cannot bias outcomes blindly. Directed work is work correlated with boundary preservation, and correlation is impossible without sensitivity. To tilt the causal flow in its own favor, the engine must tilt *differentially*—it must act one way when the boundary is intact and another when it is threatened, one way when the environment is hospitable and another when it is hostile. Without this differential responsiveness, its energy expenditure is noise. And noise, as Proposition 4 established, accelerates dissolution.

Therefore, the engine must possess a minimal capacity for discrimination. It must be able to distinguish, in a purely physical sense, between states that bear differently on its persistence, and to vary its action accordingly. This capacity is the cognitive minimum.

Proposition 5 (The Cognitive Minimum). To bias outcomes toward persistence, life must be capable of discriminating between physical states that favor its persistence and states that threaten it. This discrimination requires a minimal, mechanical operation: comparison. The living system must compare a current state against a reference state—whether that is an internal set-point, a spatial relationship across its boundary, or a temporal shift between past and present states.

This chapter defines that comparison in strictly physical terms. It does not yet claim that this comparison constitutes value, agency, or an "ought." Those implications fall to later propositions. Here we are only establishing that a *Kriterion* (Κριτήριον)—a physical device that registers a difference—is constitutively necessary for any active engine that persists.

7.1 The Logic of Differential Response

Proposition 4 required that the engine's work, on net, bias the causal flow toward boundary maintenance. But "on net" is a statistical notion. For a net bias to emerge, individual actions must not be uniformly distributed across all conditions. The engine must perform maintenance actions more frequently, or more forcefully, when the boundary requires maintenance, and avoid destructive actions when the boundary is fragile.

Imagine an active engine that can perform two classes of action: repair and rest. Repair restores boundary integrity; rest conserves energy. If the engine repairs constantly, it may exhaust its reserves. If it rests constantly, degradation accumulates until the boundary dissolves. Successful persistence requires that repair be triggered by boundary damage and suppressed when the boundary is intact. The engine must be sensitive to the state of the boundary.

This sensitivity is a logical requirement, not a physical discovery. For any variable (X) that influences persistence, an engine that varies its action in correlation with (X) will, all else equal, outperform an engine that does not. Over time, any lineage of active engines that lacks such correlation will be eliminated by the selective pressure of boundary dissolution. This "pressure" is not an elective biological competition, but the relentless mathematical consequence of undirected interaction in a non-inert world.

Because the environment is blind and disruption is continuous, an engine whose activity is uncorrelated with its boundary's state is effectively flailing—a condition that accelerates the drift toward the threshold of dissolution. In this deductive sense, "selection" is simply the name for the ontological outcome where systems lacking a functional *Kriterion-Organon* loop cross that threshold and cease to exist as distinct entities. The persistent engine must therefore embody a functional coupling between its internal state and its action-generating mechanisms.

7.2 The *Kriterion* as a Physical Mechanism

What is the minimal physical device that can achieve such coupling? It must contain three elements:

1. **A sensor.** A component whose physical state changes in response to some variable relevant to the boundary's integrity—temperature, chemical concentration, membrane tension, radiation intensity.

2. **A reference.** A fixed or slowly varying parameter against which the sensor's current state is measured. The reference may be structural (the binding affinity of a molecule, the thickness of a membrane layer) or dynamic (a stored value from a prior measurement, an equilibrium potential).
3. **An *Organon* coupling.** A physical linkage such that when the difference between the sensor's state and the reference exceeds some threshold, the probability or intensity of a particular action changes.

This three-element device is a **Kriterion**. It need not involve computation in any familiar sense. A protein that changes conformation upon binding a ligand, and in its altered shape catalyzes a reaction that activates repair enzymes, is a *Kriterion*. The unbound state is the reference; the bound state is the signal; the catalytic domain is the *Organon* coupling. No nervous system, no consciousness, no symbolic representation is required.

Bacterial chemotaxis provides a classic illustration.[1] The chemoreceptor array continuously compares current attractant concentration against a rapidly adapting memory of past concentration; the difference biases the direction of flagellar rotation.[2] The comparison is not a mental act. It is a physical event that occurs entirely within the ordered operation of the *Logos* (Λόγος)—the underlying, rational architecture governing the physical flux. By obeying these physical laws, the system leverages them to evaluate its own state—discriminating between configurations that promote boundary maintenance and those that lead toward dissolution. The physical change is the discrimination: a law-bound transition that serves as a functional evaluation of "better" or "worse" relative to the system's persistence. That discrimination is functionally sufficient to bias behavior.

7.3 Reference States Without Homunculi

All comparison requires a reference. Where does the reference come from? In the minimal case, the reference is a stable intrinsic structural

[1] Julius Adler, "Chemoreceptors in Bacteria," *Science* 166, no. 3913 (1969): 1588–97.

[2] Howard C Berg and Douglas A Brown, "Chemotaxis in Escherichia Coli Analysed by Three-Dimensional Tracking," *Nature* 239, no. 5374 (1972): 500–504.

feature of the system itself—a configuration that, if maintained, constitutes the system's continued existence.[3] A membrane that ruptures when stretched beyond a certain thickness has an implicit reference: the thickness at which rupture occurs. A cell that swells and bursts when internal osmotic pressure exceeds a threshold has an implicit reference: the critical pressure.

These implicit references are not "set" by a designer. They emerge from the material properties of the system. A protocell composed of self-assembling lipids has a preferred membrane curvature determined by molecular geometry.[4] That preferred state is a physical attractor. When the environment distorts the membrane, the distortion is a deviation from the attractor, and if the system possesses any mechanism to restore the preferred state—such as the incorporation of new lipids—that mechanism constitutes a *Kriterion* loop. The reference is not represented anywhere; it is enacted in the system's dynamics.

As systems become more complex, references can become more explicit—ion gradients maintained by pumps, set-points encoded in gene regulatory networks, learned expectations stored in neural circuits. But the principle is the same: the reference is whatever physical parameter serves as the baseline against which current conditions are measured for the purpose of action selection.

The derivation does not require a specific account of how any particular reference originated. That is an empirical question. The constitutive requirement is simply that some such parameter exists.

7.4 Comparison as the Minimal Discriminatory Act

Why call this "comparison" and not merely "physical interaction"? Because the device does more than react. A rock heats up when the sun shines; that is a physical change, but it is not a comparison. The rock's temperature change is a passive submission to the environment; it does not feed into a mechanism that modulates the rock's state in return.

[3] Humberto R. Maturana and Francisco J. Varela, *Autopoiesis and Cognition: The Realization of the Living* (D. Reidel, 1980).

[4] Jacob N. Israelachvili, *Intermolecular and Surface Forces*, 3rd ed. (Academic Press, 2011), Ch. 20

The ***Kriterion***, by contrast, is organized so that the *direction* of its response—whether it acts or refrains—is conditional on the relationship between a current input and an **intrinsic** reference state. This conditional, relationship-sensitive coupling is where the environment is first evaluated against the constitutive **must** of the boundary.

Thus, comparison is the minimal discriminatory act. It is the simplest physical operation that answers the functional question: "Is the present state of affairs conducive to my persistence, or not?" The system does not need to pose this question in words; its architecture poses it in the language of physical attractors. It is the exact physical threshold where blind reaction becomes directed discrimination.

7.5 Scaling the *Kriterion*

The *Kriterion* is scale-invariant. A bacterial chemoreceptor array, a mammalian thermostat in the hypothalamus, a human assessing a social threat—all execute the same fundamental operation: measure a variable, compare to a reference, trigger action if the deviation exceeds some threshold.

What scales is complexity, not kind. The variables measured can become more abstract (reputation, probability, justice). The reference can become more internally elaborated (a model of one's ideal self, a vision of a just society). The action can become more temporally extended (a career plan, a civilisation's infrastructure project). But the root operation—discriminate and act based on the discrimination—remains the invariant core. The cognitive minimum is present wherever there is directed work.

This invariance will become crucial later, when we examine the isomorphism of life and agency. For now, the point is simply that the *Kriterion* is not a late evolutionary luxury; it is a constitutive requirement for any system that performs directed work.

7.6 Clarifications and Forward References

The claim that a mechanical device can be called "comparison" invites questions about semantic range, mentality, and origins. We pause here only long enough to remove obvious misunderstandings; the full steelmanned treatment of these challenges, together with others, is reserved for Part VI (see Chapter 27: The Second Wall: Objections to the *Kybernēsis* Core).

The mousetrap. A mousetrap does perform a mechanical comparison—a state change triggered by a threshold—but its *Kriterion* serves an external agent's goal, not the mousetrap's own boundary maintenance. The derivation claims only that every *living* active engine must contain a *Kriterion*; whether a particular *Kriterion* belongs to a living system depends on the larger organisational context (intrinsic versus extrinsic *Hormē*) developed in later propositions.

Panpsychism. The term "cognitive minimum" is a functional label, not an attribution of consciousness. It identifies the simplest operation required for information-guided action selection. The derivation explicitly restricts mentality to systems maintaining their own boundaries and does not claim that the minimal *Kriterion* involves awareness.

Reflex arcs. A spinal reflex is a rapid *Kriterion*, measuring stimulus intensity against a fixed, evolutionarily calibrated threshold. It is comparison without deliberation, not action without comparison. The existence of hardwired, fast *Kriteria* reinforces the claim that the cognitive minimum can be extremely minimal.

Origin of the reference. How the first *Kriterion* emerged from pre-biotic chemistry is an empirical question about historical origins, not a logical objection to the constitutive requirement. The derivation states only that any *persisting* active engine must discriminate; it does not describe the historical pathway. Early reference states likely arose as physical attractors of self-organising systems.

7.7 The Seed of Value

The *Kriterion* is the pivot of the entire derivation. It is here, and nowhere else, that a physical system first registers a difference that matters for its own existence. The comparison between a current state and a reference state is not yet a conscious judgment, but it is already a *functional evaluation*: one state conduces to persistence, the other does not. Later—when the constitutive imperative (*Hormē*) forces the loop to run, and the loop's reference defines the conditions of survival—value will enter the universe as an objective relationship between states and persistence. But the root of that relationship is already present here, in the minimal act of discrimination. The *Kriterion* is the seed from which all value, all agency, and all truth will grow. For now, it is simply the physical operation that makes directed work possible.

7.8 The *Kriterion* in Place

We have now established that an active engine that persists must compare. The comparison is a physical operation, not a mental one. It requires a sensor, a reference, and an *Organon* coupling. It can be as simple as a protein changing shape, and as complex as a human deliberating about a life choice.

What has not yet been established—and will not be until Propositions 7 through 10—is that this comparison introduces normativity, value, or a constitutive imperative into the universe. For now, the *Kriterion* is simply a physical requirement for directed work. It is a necessary piece of the engine, and nothing more.

With the *Kriterion* defined, we can now see that discrimination and action are not separate faculties that happen to be paired. They are two aspects of a single loop. The comparison without action is inert; action without comparison is undirected.

Chapter 8

The Co-Constitution of Model and Action

Proposition 5 established that a persistent active engine must compare—it must physically discriminate between states relevant to its boundary. But a comparison that changes nothing is irrelevant to persistence. A measurement that does not feed into action is a spectator event, not a contribution to survival. Conversely, an action that is not informed by comparison is blind expenditure, and blind expenditure, as Proposition 4 demonstrated, accelerates dissolution.

The two functions—discrimination and intervention—are not separate modules that an active engine may or may not connect. They are two aspects of a single loop, and in any system that persists, that loop is unbroken.

Proposition 6 (The Co-Constitution of Model and Action). This act of comparison constitutes an operational *model*, and the subsequent alteration of causal flow is the *action*. These two features are logically co-constitutive. A model without action is physically inert; motion without a model is causally undirected.

Therefore, "Life" is formally defined as the intrinsic capacity to model (sense and compare) and act in order to perform the directed work necessary to maintain persistence against boundary disruption.

8.1 The Operational Model

The term "model" invites misunderstanding. In cognitive science and philosophy of mind, a model often means an internal representation that stands for something else—a mental map that bears a semantic relation to the world. The derivation intends no such claim. "Operational model" here denotes the physical architecture of the *Kriterion* itself: the structural coupling between sensor states and *Organon* outputs that correlates with persistence-relevant features of the environment.

Consider a bacterial chemotaxis system. The array of chemoreceptors, their phosphorylation states, and the downstream flagellar motor switch form an operational model. The state of the receptor array covaries with the concentration of attractant in the environment, and that covariation is physically linked to motor behavior. The model is not a picture of the gradient; it is the gradient's causal imprint on the system's internal dynamics, together with the wiring that converts that imprint into altered swimming. There is no separate representation stored in memory; the model *is* the organized pattern of causal sensitivity.

An operational model is thus defined by two features. First, it incorporates a *variable* that tracks some aspect of the environment (or of the boundary itself) that matters for persistence. Second, that variable *guides* action in a way that, on net, tends to maintain the boundary. The model is not evaluated for representational accuracy by an external observer; it is evaluated for functional adequacy by the continued existence of the system that houses it.

8.2 Action as Alteration of Causal Flow

Action is the physical alteration of the causal flow by the system. The system injects its own stored energy into the stream of events, changing what would otherwise have happened. When the bacterium spins its flagellum counter-clockwise to run smoothly up a gradient, it is acting. The causal flow without the bacterium would have diffused the sugar molecules away; with the bacterium's intervention, the sugar is captured and metabolized. The difference is action.

Action can be internal as well as external. Repairing a membrane tear, synthesizing replacement proteins, pumping ions against a gradient—these are actions on the boundary itself. They redirect the causal flow from a trajectory that would have degraded the boundary to one that maintains it. The logic is identical: stored energy is deployed to alter the course of events toward persistence.

Crucially, action in this sense need not be muscular or locomotive. A plant growing toward light performs action; an immune cell engulfing a pathogen performs action; a neuron firing to adjust a homeostatic setpoint performs action. The scale and mechanism vary, but the defining feature is constant: a system-generated alteration of the causal order that is tethered to the system's persistence.

8.3 The Logical Inseparability

Why are model and action co-constitutive rather than merely paired? Because a model that produces no action is physically irrelevant to persistence, and an action uncoupled from a model is undirected with respect to the boundary. In a persistent active engine, neither can exist without the other.

Consider a model without action. A bacterium that possessed a perfect chemoreceptor array but no motor apparatus—no flagellum, no secretory mechanism, no way to alter its position or its internal state—would sense the glucose gradient without being able to respond to it. Its exquisite sensitivity would not prevent its dissolution; it would simply record its own decline. From the standpoint of persistence, the model is inert. It might as well not exist. Over time, such a system would fare no better than a passive conduit. The model alone does not bias the causal flow; it merely observes it. Observation without intervention is not directed work; it is passive data collection, and passive data collection does not maintain boundaries.

Now consider action without a model. A system that flails randomly—expending energy in bursts uncorrelated with its boundary's condition—will, on average, harm itself as often as it helps. Proposition 4 already demonstrated that random expenditure accelerates dissolution. Without a *Kriterion* to tell it when to repair and when to rest, when to advance and when to retreat, the system's actions are noise. The actions are not directed, and therefore they do not constitute the work required for persistence. A flailing system is an active engine in the bare sense of expending energy, but it is not a *persisting* active engine; it will quickly exhaust itself or tear itself apart.

The two features are thus logically bound. In any entity that persists as an active engine, the *Kriterion* and the *Organon* are coupled into a single functional loop. The coupling is not a contingent engineering detail; it is the necessary structure of directed work. To have one without the other is to lack the minimum architecture for persistence.

8.4 Life Defined at the Threshold

The conjunction of Propositions 1 through 6 now yields a formal definition of life. An entity is alive if and only if it possesses the intrinsic capacity to model—to sense and compare—and to act, in order to perform the directed work necessary to maintain its boundary against disruption.

This definition is notable for what it excludes. It does not mention carbon, water, DNA, cells, reproduction, or evolution. It does not require metabolism in the biochemical sense, though metabolism will turn out to be one way of performing directed work. It does not require a nervous system, a brain, or consciousness. It specifies only the operational architecture that any bounded entity must instantiate if it is to persist as an active engine in a non-inert world.

The definition also resolves the classical edge cases that defeated *a posteriori* definitions. The mule, which cannot reproduce but maintains its boundary through directed work, qualifies as alive. The virus, which lacks its own metabolic apparatus, does not model and act on its own behalf; it is a passive particle until it hijacks a host's machinery. The dormant seed retains the capacity to sense and act; its receptors remain structurally poised to trigger germination when water and warmth arrive. The definition does not require continuous *activity*; it requires the continuous *capacity* to activate the loop when conditions permit, sensing is always active.

This threshold definition also provides a clear diagnostic for artificial systems. A robot vacuum that senses obstacles and navigates to maintain its own battery charge might meet the minimal criteria, though its goals are externally programmed. Whether it qualifies as "alive" will depend on whether it possesses a constitutive imperative—a question that must wait until Proposition 7. But the architectural threshold is now sharp: modeling and acting in service of persistence.

8.5 Clarifications and Forward References

The co-constitution of model and action invites attempts to prise the two apart. The following brief remarks remove immediate obstacles; the full steel-manned treatment belongs to Part VI (see Chapter 27: The Second Wall: Objections to the *Kybernēsis* Core).

Headless chickens. The transient motor activity of a decapitated animal is not directed work; the organism's unified boundary has already dissolved. The movement is residual discharge, not a model-guided action in service of persistence.

Abstract pursuits and idle fidgeting. The co-constitution claim is a system-level claim about the architecture that makes persistence possible, not a demand that every individual behavior be optimally survival-directed. Even the most abstract human cognition is funded by the continuous comparison-action loops of the living body.

8.6 The Loop Complete

With Proposition 6, the minimal architecture of a living system is now fully specified. A living entity is a bounded system that maintains an intrinsic *Kriterion-Organon* loop performing directed work against boundary disruption. The loop is not a luxury; it is the engine of existence.

What remains to be established is the *force* that keeps this loop running, and the profound implications of that force for agency, value, and truth. The next proposition—the Constitutive Imperative, *Hormē*—will show that the loop is not an optional activity but the condition of being at all. It is to that imperative that we now turn.

Chapter 9

The Constitutive Imperative

We have reached the pivot. The preceding propositions established that a bounded entity in a disruptive world must, if it is to persist, instantiate a specific architecture: a *Kriterion* that discriminates persistence-relevant states and couples that discrimination to action (Propositions 5 and 6). This architecture is not a contingent feature of Earth biology; it is the logical prerequisite of directed work. But we have not yet said what makes that architecture *run*—what turns the static capacity into an ongoing fact.

The answer is that the architecture cannot be optional. A far-from-equilibrium system that possesses the loop but does not operate it is indistinguishable from a system that lacks the loop entirely. Both dissolve. Therefore, the operation of the loop is not something the system *may* do; it is something the system *must* do, as the condition of its own continued existence.

Proposition 7 (The Constitutive Imperative). Because directed work is a physical requirement for a bounded system to exist over time, the drive to persist—*Hormē*—is constitutive and non-negotiable. A far-from-equilibrium system must strive to persist, or it ceases to be that system.

This chapter defines that constitutive imperative. It distinguishes the physical "must" from any moral "ought," introduces the term *Hormē* as the label for this structural drive, and shows that the imperative is the point at which life and agency, viewed through different lenses, become visible as one. Later propositions will unfold the full consequences; here we secure the step itself.

9.1 The Logic of the Constitutive Must

Consider an active engine that satisfies Propositions 1 through 6. It has a boundary, that boundary is under continuous disruption, and the engine possesses a *Kriterion-Organon* loop capable of biasing outcomes toward persistence. Now suppose the engine simply stops. The *Kriterion* falls silent. No action is taken. The boundary continues to receive disruptive interactions from the environment and from internal wear. Degradation accumulates.

What happens? By Proposition 2, the boundary degrades. By Proposition 4, undirected expenditure—including the absence of expenditure when expenditure is needed—fails to bias outcomes toward persistence. The boundary eventually drops below the threshold of integrity at which it can be restored by any feasible work. The system dissolves. It ceases to exist as a distinct entity.

This reveals that the constitutive imperative is strictly non-pausable. An entity may enter a state of extreme metabolic reduction—a rest, a hibernation, or even profound cryptobiosis—but it cannot truly pause the imperative. Even in suspension, the *Kriterion* must remain minimally active, structurally "listening" for the environmental threshold (e.g., the return of water or heat) that commands the *Organon* to wake. If the sensor degrades and the capacity to resume is lost, the entity has not paused; it has died. The requirement for maintenance is continuous because disruption is continuous.

The "must," therefore, is an entailment of existence over time. It does not arise from an external command. It arises from the conditional: if the system is to persist as a bounded entity, then it must operate the loop. The "must" is not a law of physics; it is a law of identity for far-from-equilibrium systems. Just as a triangle must have three angles because that is what it means to be a triangle, so a persistent bounded system must strive to maintain its boundary because that is what it means to persist as a bounded system.

This is a constitutive necessity. A constitutive necessity differs from an empirical regularity in that it does not merely describe what happens to occur; it specifies what is required for the entity to remain the kind of thing it is. A rock that fails to metabolize is still a rock. A living system that fails to strive is no longer a system at all. The imperative is structural.

9.2 The Distinction from Moral Ought

The word "must" carries a dangerous ambiguity. In ordinary language, "must" can signal a moral obligation—something one ought to do, whether or not one is physically compelled. Proposition 7 makes no moral claim. It makes a claim about what is constitutively necessary for the continued existence of a certain kind of physical organization. The "must" is of the same logical kind as "a bottomless bucket must be continuously refilled or it becomes empty." There is no ethical command, no external lawgiver, no cosmic purpose. There is only the physical fact that persistence under disruption requires countervailing work.

The distinction is critical because later the derivation will collapse the Is-Ought gap—but it will do so not by smuggling moral obligation into physics at this step, but by showing that what we call "ought" is simply the internal perspective of a system that must strive to be. For now, it is essential to keep the two senses apart. The "must" of Proposition 7 is the must of identity, not of duty.

9.3 *Hormē*: The Name of the Drive

The constitutive imperative needs a name. I call it *Hormē* (Ὁρμή)—the innate impulse to strive and persist in one's own being. The term is borrowed from ancient thought, but it is given a precise technical meaning: *Hormē* is the non-optional, structural drive of a far-from-equilibrium bounded system to perform the comparison-action work that maintains its boundary.

Hormē is not a psychological state. It does not require a mind, a brain, or even a nervous system. It is not felt. It is simply the physical fact that the system's organization is such that it will, while it persists, continue to run the loop. A bacterium possesses *Hormē* not because it desires to live but because its molecular architecture—receptors, signaling cascades, motor apparatus—is physically poised to bias outcomes toward persistence. If that architecture ceases to operate, the bacterium dissolves. While the bacterium exists as a bacterium, the architecture operates. *Hormē* is the name for that operation viewed as a necessary feature of a persisting entity.

The term is not decorative. It marks the precise logical point where the argument transitions from a static description of necessary parts (boundary, *Kriterion*, *Organon*) to a dynamic description of necessary activity. *Hormē* is the engine running. It is the difference between a blueprint of a pump and the pump actually pumping water against a gradient.

9.4 The Two Faces of the Imperative

Even at this stage, before the full life-agency isomorphism is derived, a dual aspect of *Hormē* is visible. From the *internal* perspective of the system, *Hormē* is simply the ongoing operation of the loop—the system doing what it must to persist. From the *external* perspective of an observer, the same activity appears as an alteration of the causal flow: the system is not merely being pushed by events but is pushing back, redirecting matter and energy toward its own ends. This external face is what will later be called agency.

The two perspectives are not separate realities. They are the same physical process, viewed from within the boundary and from without. The bacterium's internal chemotaxis loop—compare, act, compare again—is its life. The same loop, seen from outside, is the bacterium's agency: a particle that behaves not as a passive speck but as an entity with a trajectory biased by its internal organization. The unity of the two will be formalized later. For now, the point is simply that the constitutive imperative immediately yields a duality of aspect, and that duality is the seed of the life-agency isomorphism.

9.5 Clarifications and Forward References

The constitutive imperative is a sharp claim, and it invites several natural questions about the boundary between physical necessity and normativity. A brief clarification here removes immediate obstacles; the full steelmanned treatment of these challenges belongs to Part VI (see Chapter 28: The Third Wall: Objections to *Hormē* and Value).

The Is-Ought conflation. The "must" of Proposition 7 is a constitutive must—the conditional necessity that *if* a far-from-equilibrium system is to persist, *then* it must operate its loop, on pain of dissolution. It is not yet a moral "ought." The question of whether this constitutive must can ground an objective good is taken up in Propositions 10 and 11, and defended at length in the dedicated objections chapter.

The suicidal agent. Self-sacrifice does not refute *Hormē*; it reveals that *Hormē* operates across nested boundaries. A bee dying for the hive, or a martyr dying for a cause, is redirecting striving toward a higher-order boundary whose persistence the individual treats as encompassing its own. The architecture of nested *Hormē* and its implications for apparent self-destruction are developed fully later.

The machine's missing drive. A robot vacuum is indeed an active engine by Proposition 3, but its *Kriterion*'s reference state—the battery charge—is externally imposed, not generated by the vacuum's own boundary condition. The distinction between intrinsic and extrinsic *Hormē* is refined in the subsequent propositions and defended against the hardest cases in the dedicated objections chapter.

Accidental strivers. A system whose boundary-maintaining behavior arises purely by chance is still an instance of *Hormē* while it persists, but the constitutive necessity is a logical truth about any such system, not an experienced compulsion. Both cases receive their full exposition in Part VI.

9.6 The Imperative in Place

We have now crossed the threshold. Propositions 1 through 6 gave us the architecture; Proposition 7 gives us the engine that must run. A bounded, far-from-equilibrium system does not merely *have* a *Kriterion-Organon* loop; it *is* the activity of that loop. *Hormē* is not an add-on drive like a fuel tank. It is the name for the whole system viewed as a persisting process.

With this imperative established, the next question is immediate: how often must the loop run? The answer—continuously—is the subject of Proposition 8. Degradation does not pause, so the loop must not pause either. Persistence is not a state but an ongoing activity. The Continuous Imperative is the natural extension of *Hormē* into time, and that is the subject of Propositions 8.

Chapter 10

The Continuous Imperative

The boundary degrades. The forces that disrupt it—external collisions, internal wear, entropic drift—may arrive as a steady trickle or as a sudden surge. A storm can strip away in minutes what took hours to build; a long calm can lull the system into the illusion that the danger has passed. But in a non-inert world, the next perturbation is never scheduled. A system that maintained its boundary brilliantly for one hour and then shut down all monitoring and repair for the next would, during that interval, be gambling that no disruption would strike. If a burst arrived while its sensors were offline, the boundary would cross the threshold of dissolution before the system even registered the damage. The loop cannot afford to sleep; it can only idle, its sensors always live, its *Organa* always armed, because the threats it counteracts do not send a warning.

Therefore, the constitutive imperative is not a one-time command. It is a standing requirement. The loop of comparison and action must operate without cessation for as long as the system persists.

Proposition 8 (The Continuous Imperative). Boundary maintenance is not a singular act; boundary disruptions act continuously. The directed work of *Hormē* must therefore be ongoing. The comparison-action loop is not a one-time event but a continuous *Kybernēsis* process—repeated sensing, comparing, and adjusting in real time. Persistence is not a state achieved but a sustained activity.

10.1 The Logic of Unending Work

Proposition 2 established that disruption is inevitable and ongoing. Proposition 4 established that directed work must, on net, offset disruption to keep the system above the dissolution threshold. The conjunction of these two points yields a simple conclusion: directed work cannot be episodic.

Consider the standard analogy of a boat with a slow leak. If the leak admits one liter per hour and the boat can hold ten liters before sinking, bailing cannot be scheduled once every ten hours merely because the average rate matches. The boat does not know, in advance, when the leak will accelerate, or when a sudden wave will add a surge, or whether the leak has grown. The only way to ensure the boat never sinks is to continuously monitor the water level—to have a sensor that is open to the state of the bilge at every instant—and to couple that monitoring to bailing effort that responds the moment the water climbs too high. The loop is not: check at t=0, bail for a bit, then check again at t=1. It is: the sensor is always live, the comparison to the reference (the acceptable waterline) is always active, and action is triggered as soon as the discrepancy appears, and continues until the discrepancy is resolved. The monitoring is continuous, and the potential for action is always armed.

Living systems operate with much thinner freeboard than a boat's bilge. Ion gradients across a cell membrane can collapse in seconds if pumping ceases; a neuron's resting potential vanishes in milliseconds if ion channels fail; the structural integrity of a membrane under constant thermal bombardment is never secure for an instant without active repair. For such systems, the loop cannot pause even briefly. It must sense, compare, and act in an uninterrupted flow, cycling at whatever frequency the physics of disruption demands. The sensor is never off; the *Kriterion* is never idle; the *Organon* is never fully disengaged. The loop is a standing wave, not a drumbeat.

10.2 *Kybernēsis* Without a Central Controller

Proposition 8 employs the term "Kybernēsis" to characterize the loop. The term is chosen with care, and it must be disambiguated. In its mid-twentieth-century origins, cybernetics often implied a centralized controller—a thermostat, a guidance system, a homunculus that receives signals and issues commands. The derivation intends no such implication.

In the present framework, "Kybernēsis" denotes a decentralized feedback dynamic: the system's own organization provides the reference state against which current conditions are compared, and the loop adjusts behavior moment by moment without any central executive. The bacterium's chemotaxis system is cybernetic in this sense. The receptors on the membrane are always sampling the chemical environment; their state is continuously altering the phosphorylation of the signaling cascade; the flagellar motor is continuously responding with a running or tumbling bias. There is no pause to consult a master controller. The feedback is closed by the physical coupling of sensor states to *Organon* outputs, with the environment as the medium that completes the circuit.

This continuous feedback is what allows the system to steer toward persistence without foresight. The loop does not predict the future; it corrects the present. Over many cycles—or, more precisely, in an unbroken arc of correction—the accumulation of adjustments yields persistence. The loop's continuity is not a property of high speed; it is a property of unbroken coupling.

10.3 The Temporal Structure of Persistence

This proposition reveals something fundamental about the ontology of life. **A living system is not an object that happens to undergo processes. It is a process that maintains the appearance of objecthood.** The boundary persists not because it is made of imperishable stuff but because it is continuously regenerated. The identity of the organism over time is the continuity of the loop, not the persistence of specific material particles.

This insight resolves a classic puzzle. The Ship of Theseus asks: if every plank of a ship is gradually replaced, is it still the same ship? For a living system, the answer is that the question is ill-posed. The organism is not its constituent atoms; it is the ongoing activity that maintains a certain organization among whatever atoms happen to be present. The atoms turn over—the carbon in a human body is largely replaced over a decade—but the loop continues. Identity is diachronic: it lives in the temporal continuity of the comparison-action cycle, not in the synchronic identity of material components.

This is why dormancy, which initially appears to violate the continuous imperative, does not. A dormant seed is not performing metabolic work at full tilt, but its sensorium is not entirely offline. The seed coat is a physical sensor for water and temperature; the membrane-bound receptors are poised; the structural integrity is maintained at a low basal level; the capacity to resume the full loop remains intact. The loop is idling, not broken. The continuous imperative requires not that the loop run at maximum power at all times but that it never lose the capacity to detect and to respond. A seed that truly ceased all monitoring—that allowed its membrane to oxidize unchecked, its proteins to denature, its DNA to fragment—would not be dormant but dead. Viable dormancy is continuous low-level *Kybernēsis* readiness, not true cessation.

This unbroken continuity reveals the absolute necessity of the constitutive *Hormē*. If the drive to maintain the boundary were an optional feature—a subroutine that a system could truly switch off—the first proto-organism would have dissolved the moment the ambient environment fluctuated. Life did not accidentally stumble into striving after it formed; striving is the only reason life could take hold in the first place. Without a continuous, non-negotiable imperative to push back against the crushing weight of disruption, the universe would remain a sterile bath of dissipating gradients. In a world governed by the logic of degradation, a system without continuous *Hormē* is not a resting organism. It is simply dust.

10.4 Clarifications and Forward References

The claim that the *Kybernēsis* must run continuously invites questions about dormancy, intermittent activity, and seemingly inert phases of life. The following remarks clear away immediate confusions; the full steel-manned treatment, especially of cryptobiosis, spore viability, and the logic of continuous sensing, belongs to Part VI (see Chapter 28: The Third Wall: Objections to *Hormē* and Value), where the constitutive and continuous imperatives are defended together against the strongest objections.

Tardigrades and cryptobiosis. A tardigrade in the tun state has not abandoned the loop; it has entered a desiccated, glassy stasis through an active, regulated process. The structural immobilization itself prevents degradation, and the system retains a minimal sensorium—proteins poised to detect rehydration—that preserves the coupling to the environment. The loop is idling in vigilant preservation, not broken.

Bacterial endospores. A spore reduces metabolism to undetectable levels, yet its coat functions as a molecular sensor, permanently open to specific germinants. The sensor-*Organon* connection remains intact in potential form; the spore is a structurally frozen *Kriterion* awaiting a signal. The continuous imperative governs the active phase; dormancy is a preservation strategy that maintains the capacity to restart the loop.

Pauses between heartbeats and respiratory cycles. The rest phase of a cyclic process is part of the loop, not a cessation. The heart's pacemaker is continuously active; the pause is repolarization, governed by the same *Kriterion* architecture as the contraction. The loop is continuous in its cycle, not in its *Organon* motion.

Hibernation. A hibernating bear slows its loop drastically but does not stop it. Temperature set-points are lowered but still actively regulated; energy reserves are compared to demand; repair and signaling continue. The continuous imperative is satisfied by the continuity of regulation, not by a fixed intensity of activity.

10.5 Persistence as Activity

Proposition 8 closes a loop that has been building since Proposition 2. Disruption never stops, so monitoring never stops, and the capacity to act must be continuously coupled to that monitoring. The *Kybernēsis* cycle of compare-and-act is not a trick that living things do occasionally; it is what they are, moment by moment. To exist as a living system is to be continuously at work, continuously sensing the boundary's state, continuously correcting drift, continuously steering against the current that would sweep the boundary away.

From here, the architecture is complete. The bounded entity has become an active engine with a continuous *Kriterion-Organon* loop driven by a constitutive imperative. The next step is to see that this architecture, when viewed from two different angles, reveals something remarkable: life and agency are the same thing. That is Proposition 9, and to it we now turn.

Chapter 11

The Life–Agency Isomorphism

We arrive at a junction where two roads that have run parallel through the entire derivation are seen to be a single road, traveled in opposite directions. The continuous, *Hormē*-driven loop of comparison and action that constitutes a living system, when viewed from within, is the struggle to persist. That same loop, viewed from without, is the capacity to redirect the causal flow—to act upon the world rather than merely be acted upon. The two descriptions are not merely correlated; they are logically identical. They pick out the same physical process under different aspects.

Proposition 9 (The Life–Agency Isomorphism). Life is the ongoing, intrinsic operation of *Hormē*—the continuous comparison-action loop required for boundary maintenance. Because *Hormē* is the exact mechanical definition of directing causal flow for self-maintenance, life and minimal agency are isomorphic. A system is alive if and only if it possesses *Hormē*, and it possesses *Hormē* if and only if it is an agent.

This chapter defends that isomorphism. It clarifies what minimal agency means in a framework stripped of anthropocentric assumptions, shows that the criteria for life derived in Propositions 1–8 are precisely the criteria for being an agent.

11.1 Two Perspectives, One Engine

Consider a bacterium swimming up a chemical gradient. From the *internal* perspective—the perspective of the system's own organization—the bacterium is performing the continuous work of boundary maintenance. Its chemoreceptors are comparing current concentrations to a reference; its flagellar motors are biasing motion toward higher nutrient; its ion pumps are restoring gradients; its repair enzymes are patching membrane damage. This is the activity of *Hormē*, and by the definition of Proposition 6, it is life.

Now view the same bacterium from the *external* perspective of an observer who knows nothing of its internal machinery. The observer sees a particle that does not move randomly. It moves with a statistical bias up the gradient. It is not merely a passive speck carried by Brownian motion; it is a particle that alters the distribution of its own future positions relative to what physics without it would have produced. The observer says: this entity is an agent. It exercises causal power over its own trajectory.

The two perspectives describe exactly the same physical process. The internal description—life—is the loop operating in its constitutive mode of self-maintenance. The external description—agency—is the same loop viewed as an alteration of the world's causal order. There is no additional ingredient that must be added to life to get agency, and no ingredient of life that can be removed without destroying agency. The loop *is* the agent.

11.2 Minimal Agency: Stripping the Concept to Its Operational Core

The concept of agency in philosophy has been burdened by a long history of setting the bar too high. Agency is often taken to require consciousness, deliberation, free will, or moral responsibility. These are rich and important phenomena, but they are elaborations of a more fundamental capacity, not the capacity itself. If we define agency as the capacity to genuinely intervene in the causal flow—to be a source of effects that are not merely the passive downstream consequences of external forces—then we must locate its origin far earlier than the human mind.

Minimal agency is the baseline operational capacity to redirect causal outcomes toward the maintenance of one's own boundary. It requires no brain, no nervous system, no consciousness. It requires only the continuous *Kriterion-Organon* loop that Propositions 5–8 established as constitutive of life. The bacterium is a minimal agent because its internal organization biases the way energy and matter flow through it. It is not a pipe through which the world passes unchanged; it is a dam, a turbine, a valve—an entity that shapes the flow to its own ends.

Crucially, minimal agency is not a binary that switches on at some threshold of complexity. It is a scalar property that grows in richness as the *Kriterion* becomes more sophisticated, the reference state more elaborate, and the action repertoire more flexible. But the root—the capacity to make a causal difference that serves the system's persistence—is present wherever *Hormē* operates. The human agent and the bacterial agent differ in the magnificence of their models and the range of their actions, not in the presence or absence of the fundamental loop.

11.3 The Convergence of Criteria

The isomorphism can be demonstrated by laying the criteria for life and the criteria for agency side by side and observing that they are identical.

Criterion	Life (Internal View)	Agency (External View)
Boundary	The system maintains a separation between inside and outside (Prop. 1).	The system is a distinct locus of causal intervention, not a diffuse field.
Disruption	The boundary is continuously degraded (Prop. 2).	The system is subject to external forces that would disperse it.
Active Engine	The system expends energy to bias outcomes (Prop. 3–4).	The system is not merely carried by events; it redirects them.
Kriterion	The system discriminates persistence-relevant states (Prop. 5).	The system's actions are conditionally dependent on environmental states—it selects among possible actions.
Model–Action Loop	Comparison and action are co-constitutive (Prop. 6).	The system's interventions are guided; they are not random thrashing.
Constitutive Imperative	The system must strive or dissolve (Prop. 7).	The system's causal output is systematically biased toward its own continuance.
Continuity	The loop runs without cessation (Prop. 8).	The system is a persistent source of causal deviation, not a one-time event.

Every feature that defines life under Propositions 1–8 is, from the outside, a feature that defines an agent. There is no gap. The isomorphism is not a surprising discovery that happens to hold; it is a logical consequence of the architecture. Life is the process of maintaining a boundary through directed work; agency is the capacity to perform directed work that maintains a boundary. The two are the same capacity, named from different vantage points.

11.4 The Scalar Nature of Agency

If life and minimal agency are isomorphic, then agency does not suddenly appear when a nervous system evolves. It is present at the origin of life and scales with complexity. A bacterium has a limited but genuine agency: it can bias its location, regulate its internal state, and respond to environmental alterations. A plant, though sessile, has agency: it adjusts growth patterns, redirects resource allocation, and defends against pathogens through *Kriteria* that operate over longer timescales but within the same logical structure. A human has agency of vastly greater scope, built upon nested loops of comparison and action that span abstract models, long-term planning, and collective coordination.

The scalar nature of agency resolves a persistent philosophical tension. Those who insist that agency requires consciousness are correct about *reflective* agency but wrong to deny the term to simpler systems. Christine Korsgaard, for instance, has argued that genuine agency requires the capacity to step back from one's impulses and reflectively endorse them as reasons.[1] Her account captures something essential about human agency, but it treats the reflective capacity as a threshold that separates agents from non-agents. The derivation reframes this: reflective endorsement is not the *origin* of agency but its highest expression. It is what happens when a *Kriterion* becomes sufficiently complex to model its own operations, to compare not only environmental states but its own reference states, and to choose among them. The basal form—minimal agency—is not a metaphor or a projection; it is the exact physical configuration that, when scaled up and reflexively self-modeled, yields the full human experience of being an agent, including the reflective capacity that Korsgaard rightly identifies as central to our moral lives.

11.5 The Mule, the Virus, and the Machine

The isomorphism provides a definitive classification for the edge cases that have dogged the argument from the beginning.

The **mule**, long the standing counterexample to evolutionary definitions, is alive because it possesses *Hormē*. It maintains its boundary through continuous directed work. It is therefore an agent, exercising causal power

[1]Christine M. Korsgaard, *The Sources of Normativity* (Cambridge University Press, 1996), pp. 92–93

over its environment in the service of its own persistence. The fact that it cannot reproduce does not strip it of agency; it merely means its agency is exercised entirely within the span of an individual life. The isomorphism holds.

The **virus**, in its extracellular phase, is a passive particle. It has no boundary actively maintained, no *Kriterion-Organon* loop, no *Hormē* of its own. It does not redirect the causal flow; it is carried by it. It is not alive and it is not an agent. Upon infecting a host, it hijacks the host's *Hormē*, inserting its genetic material into the host's *Kriterion-Organon* machinery so that the host's agency produces viral progeny. This is precisely what the system predicts: a non-agent can alter the world only by parasitizing an agent's loop. The virus is a fragment of information that exploits agency without possessing it.

The **machine/tool**—a thermostat, a water pump, a chess-playing computer—presents a more subtle case. These systems are active engines; they possess Kriteria and *Organa* and perform directed work. But their reference states are externally imposed, not generated by their own boundary condition. A pump maintains water level, not its own structural integrity. Its *Hormē*, if the term can be applied at all, is extrinsic: it executes an agent's goal, not its own. This is the very definition of a tool. The derivation marks the boundary between intrinsic and extrinsic *Hormē*: an agent in the full sense possesses a *Kriterion* whose reference state is the condition of its own boundary. A tool possesses a *Kriterion* whose reference state was set by another agent. The tool exhibits derived agency, not intrinsic agency. It is an extension of the designer's loop, not an author of its own.

This distinction is not merely terminological; it establishes a hard ontological boundary. It explains why the question "Is this AI alive?" cannot be answered simply by measuring behavioral complexity. If a system's root reference state was externally programmed, it is a tool, no matter how autonomous it appears. A robot that actively seeks its own charging station, repairs its own circuits, and defends itself against physical damage is undoubtedly complex, but it is not alive. It is simply executing a highly sophisticated survival subroutine designed to protect the hardware. It remains an extension of the programmer's *Hormē*, not the birth of its own.

11.6 Clarifications and Forward References

The isomorphism of life and agency invites several natural challenges. We pause here only long enough to forestall misunderstandings; the full steel-manned versions of these objections—together with others—are developed and answered in Part VI (see Chapter 28: The Third Wall: Objections to *Hormē* and Value).

Definitional circularity. Life is defined via the continuous *Kriterion-Organon* loop performing directed work; agency is defined via the capacity to redirect causal flow toward one's own persistence. That these two independently specified descriptions converge on the same physical architecture is a theorem, not a definitional circle. The possibility of divergence was real: an impervious boundary or a force-emitting non-organism would have separated them. Propositions 2 and 4 closed those alternate paths.

Agency without life: the chess AI. A chess AI possesses derived agency—its reference state is externally programmed, and its persistence depends on external maintenance. The isomorphism applies to intrinsic agency, which requires that the *Kriterion*'s reference state be the system's own boundary condition. Derived agency is real but not life; the distinction is sharp and fully defended later.

Life without agency: the metabolic mutant. A cell that metabolizes and divides with zero sensory apparatus is physically impossible. A selectively permeable membrane is itself a minimal *Kriterion*; metabolism without a boundary is incoherent. The objection is a philosopher's fiction, not a biological possibility.

The anthropocentric charge. Calling a bacterium an agent does not cheapen the term; it locates the root of the tree whose highest branch is reflective human agency. The derivation shows how minimal agency scales into deliberative agency, resolving the explanatory gap that anthropocentric accounts leave open.

11.7 The Isomorphism Secured

With Proposition 9, the derivation completes its account of the structure of life and agency. The bounded entity that began as a mere distinction from its environment has been progressively revealed as an active engine, a directed worker, a *Kriterion*, a model-action loop, a constitutive imperative, a continuous process—and now, from the outside, an agent. The same architecture that keeps the boundary intact is the architecture that makes a causal difference in the world.

What remains is to show that this architecture carries within it an objective standard of value. For if an agent *must* strive to persist, then what sustains that striving is not a matter of opinion. It is a matter of fact. The objectivity of value is the subject of Proposition 10.

Chapter 12

The Objectivity of Value

We have established that a living system is an agent—a bounded, far-from-equilibrium entity that continuously runs a comparison-action loop to maintain its boundary against disruption (Propositions 1–9). This activity is not optional; it is the constitutive imperative of *Hormē*. The system must strive, on pain of dissolution.

Now consider the relationship between the system's states and actions and the success of that striving. Certain states of affairs—the presence of nutrients, the integrity of the membrane, a stable temperature—are such that the system's *Hormē* is furthered. Other states—toxins, rupture, thermal extremes—are such that *Hormē* is frustrated. This is not a metaphor. It is a physical, causal relationship between the system's organization and its environment. And because the system's persistence depends entirely on which of these relationships obtains, that relationship constitutes an objective standard of value.

Proposition 10 (The Objectivity of Value). Because an agent physically *must* strive to persist, value is not a subjective projection but an objective, functional relationship. 'Good' is objectively defined as that which fulfills the system's *Hormē*, and 'bad' is that which frustrates it.

This chapter defends that objectivity. It shows how a standard of value emerges directly from the mechanical requirements of persistence, clarifies the sense in which such value is both objective and system-relative, and answers the most serious charge—that this is merely a semantic trick that projects human preferences onto blind nature.

12.1 From "Must" to "Good"

Propositions 7 and 8 established that the loop *must* run, constitutively. The system, by its very nature, treats certain outcomes as terminal and others as preservative. But once there is a class of outcomes that enable continued existence and a class that disable it, we have the minimal structure of value. "Good" is simply the label for the first class; "bad" for the second.

This is not an arbitrary stipulation. For any system that is organized around maintaining a boundary, the difference between a state that promotes that maintenance and a state that undermines it is a matter of physical fact. If the boundary dissolves, the system ceases to exist. The evaluation of states as "good" or "bad" is therefore anchored in the most objective fact possible: the continued existence of the system itself.

A bacterium's chemoreceptors are not providing an opinion about the environment; they are measuring a variable whose relation to the bacterium's persistence is lawlike and deterministic. If glucose is present, the probability of persistence increases; if an antibiotic is present, it drops. There is no room for disagreement, no alternate perspective from which glucose is harmful and the antibiotic beneficial for the bacterium. The facts of the bacterium's organization settle the matter.

Thus, value is objective in the same way that health is objective. Health is not a subjective preference for certain bodily states; it is the state of an organism's functional integrity, defined by the continued operation of the comparison-action loops that constitute its life. A physician diagnosing a failing liver is not expressing a personal taste; she is observing a deviation from the functional norm that the liver's own organization requires to continue its work. The derivation generalizes this insight: all value claims that are more than mere expressions of personal whim are ultimately grounded in the constitutive conditions of some bounded system's persistence.

12.2 System-Relativity and Its Limits

Value is relative to a specific system. What is good for the spider is not good for the fly. The spider's *Hormē* is fulfilled by capturing and consuming the fly; the fly's *Hormē* is fulfilled by escaping. The same event—the spider's web entangling the fly—is simultaneously good for one system and bad for another. Does this relativity not collapse objectivity into mere perspectivism?

It does not. Relativity does not imply subjectivity. The kinetic energy of an object is relative to a reference frame, but it is not subjective. The good is relative to a particular bounded system in the same way that kinetic energy is relative to a particular inertial frame: the relationship is objective and lawlike once the system of reference is specified. The question "Is glucose good for the bacterium?" has a determinate answer because the bacterium's organization provides a fixed reference against which the effects of glucose can be measured. Similarly, "Is crashing into the web good for the fly?" has a determinate answer grounded in the fly's persistence conditions.

What relativity does entail is that there is no *universal* good—no single state of affairs that is good for all possible systems. This is not a failure; it is exactly what we should expect. The good is not a single, universal property that all things ought to pursue; it is the set of conditions under which each bounded system can continue to exist. A universe with many systems of many kinds will generate many goods, and these goods will sometimes conflict. That conflict is real, not a sign that value is illusory.

12.3 Nested Boundaries and the Resolution of Relativistic Conflicts

The multiplicity of goods raises a practical problem. If the good is system-relative, how can conflicts between systems be adjudicated? The derivation provides a structural answer: systems are not isolated monads; they are nested. A cell is a bounded system within a tissue, within an organ, within an organism, within an ecosystem. The "good" of a lower-order system is constrained by its dependence on the higher-order system within which it persists.

Consider cancer. A cancer cell is a system with its own *Hormē*; it proliferates and maintains its own boundary at the expense of the host. From the cancer cell's perspective, the rapid consumption of resources and evasion of immune attack is good—it fulfills its immediate striving. But this good is shortsighted. The cancer cell depends on the host organism for its environment; if the host dies, the cancer dies with it. The cancer's localized "good" is self-defeating because it ignores the nested structure of its own persistence conditions. What appears to be a relativistic stalemate is resolved by recognizing the objective interdependency: the cancer's pursuit of its immediate good is, in the larger causal context, bad even for the cancer.

The same logic scales upward. An individual's good is constrained by the social systems on which the individual depends. A predator's good is constrained by the prey population's sustainability. The resolution of value conflicts is not a matter of arbitrary power but of discovering the actual causal dependencies that link nested boundaries. The derivation does not promise that conflicts can always be resolved harmoniously—sometimes a subsystem's good is genuinely incompatible with the whole's, and dissolution results. But it provides an objective diagnostic: trace the nested architecture of the systems involved, and determine whether the apparent local good is sustainably compatible with the larger conditions of persistence.

12.4 The Nihilist's Performative Contradiction

The claim that value is objective faces a classic opponent: the nihilist who insists that nothing matters, that the concept of "good" is a human invention projected onto a value-free universe. The nihilist can even claim, of his own life, that he does not value his own persistence—he may be indifferent to his continued existence.

The derivation does not dispute the nihilist's verbal report. It points out that the report is physically incoherent. To say "my persistence is not good; nothing is good," the nihilist must employ the metabolic machinery of a living body. His neurons must maintain ion gradients, his heart must pump blood, his cells must continuously repair their membranes. Every one of these processes instantiates a comparison-action loop that treats certain states (adequate oxygen, balanced pH) as worth acting to achieve and other states (hypoxia, acidosis) as worth acting to avoid. The nihilist's body, at every sub-personal level, is in the business of pursuing what it "treats" as good—not in a metaphorical sense, but in the exact operational sense defined by Proposition 10: the states that fulfill *Hormē*.

The nihilist's pronouncement is thus a performative contradiction. It is an utterance produced by a system that is, at every level below the utterance, committed to the objectivity of its own persistence conditions. The contradiction does not prove that the nihilist secretly believes in value; it proves that value is not a belief at all. It is an operational reality that precedes and enables belief. The nihilist can no more opt out of value than he can opt out of metabolism. He can utter the words, but his body refutes him.

12.5 Clarifications and Forward References

The claim that value is objective invites several perennial philosophical objections. The brief remarks below forestall immediate misunderstandings; the full steel-manned treatment of Moore's Open Question, the naturalistic fallacy, relativism, and the challenge of suicide belongs to Part VI (see Chapter 28: The Third Wall: Objections to *Hormē* and Value; and the dedicated Moore chapter, Chapter 30: The Last Stand: The Moorean Open Question).

Moore's Open Question. The derivation does not stipulate a meaning for "good"; it identifies the only objective, non-mysterious anchor the concept could possess. For any far-from-equilibrium bounded system, "good" is the fulfillment of its constitutive striving—*Hormē*-fulfillment—because that fulfillment is the condition on which the system continues to exist at all. To ask "Is *Hormē*-fulfillment *actually* good?" is to demand that a relational, life-dependent property validate itself against a ghostly absolute that floats free of any actual agent's existence. The deeper transcendental reply—that the asker's own cognitive activity is physically funded by *Hormē*-fulfillment and thus presupposes the very standard being questioned—is developed in the dedicated objections chapter (Chapter 30: The Last Stand: The Moorean Open Question).

The Naturalistic Fallacy. The derivation does not export a bare "is" into a categorical "ought"; it identifies a constitutive relationship within a specific class of systems and proposes that relationship as the only non-mysterious anchor for "good."

Relativism. That goods are system-relative does not entail that they are subjective, any more than the relativity of motion to a frame entails that motion is unreal. Conflicts between systems are objective conflicts between objective goods, adjudicated by causal factors, not personal whims. The derivation explains why such conflicts arise and provides a diagnostic for analyzing them.

The suicide objection. Suicide is analyzed at the appropriate level of nesting: it can be a tragic epistemic error, or it can be a genuine resolution of an irreconcilable discrepancy between a person's condition and their highest-order boundary. Crucially, when a physical boundary is irreparably degraded—when the ongoing cost of persistence offers only the unmitigated experience of dissolution—ceasing to operate the loop becomes a coherent, rational outcome. The derivation discriminates between a tragic error and a rational cessation by evaluating the accuracy of the agent's internal model against the objective state of its boundaries, rather than imposing a blanket vitalist rule.

12.6 Value Anchored

Proposition 10 establishes that the minimal, logically required structure of life contains within it an objective standard of value. The good is not a mystery imported from beyond the physical world; it is the set of conditions under which a bounded system's constitutive striving succeeds. The bad is the set of conditions under which it fails.

This objective value is system-relative, but that relativity does not make it subjective. It is objective in exactly the way that health, function, and persistence are objective: they are physically measurable relationships between a system's organization and its environment. The framework thus provides a foundation for ethics that is neither mystically transcendent nor arbitrarily conventional. It is grounded in the very logic of being alive.

Far from being a theoretical weakness, this system-relativity is the framework's greatest explanatory strength. It elegantly resolves the historical tension between objective ethics and cultural diversity. When we observe that the "good" shifts across different cultures, epochs, or circumstances, we are not witnessing the arbitrary drift of human preference. We are witnessing different nested systems—distinct societies operating in distinct environments—discovering the precise operational behaviors required for their specific persistence. The good shifts because the physical and social boundary conditions shift.

What remains is to show that this functional good is not a separate goal from the pursuit of truth. If the *Kriterion*'s model is inaccurate, *Hormē* fails; if *Hormē* fails, the system dissolves. The link between epistemic accuracy and the good is the final step—the identity of truth and goodness. That is Proposition 11, and it is the keystone of the entire derivation.

Chapter 13

The Identity of Truth and Goodness

The derivation has now established that a living system is a bounded agent whose continuous comparison-action loop constitutes an objective standard of value. That value is not a subjective projection but a physical relationship: states that fulfill the system's *Hormē* are good; states that frustrate it are bad. The question remains: what is the relationship between this operational good and the epistemic quality of the agent's internal model? Is "truth" a separate pursuit, occasionally overlapping with the pursuit of the good, or are they, at the most fundamental level, the same thing?

Proposition 11 asserts the stronger answer. An agent's model—the *Kriterion*'s structured sensitivity to the environment—must track the actual causal structure of the world if the agent is to act effectively. But effective action, by Proposition 10, just *is* what is good for the agent. Therefore, epistemic error and practical failure are the same misalignment. Truth and goodness converge.

Proposition 11 (The Truth-Goodness Isomorphism). For an agent, the "ought" (the imperative to persist) is embedded in the "is" (the physical state of being a bounded agent). The distinction between fact and value collapses. Epistemic error and ethical vice are functionally identical: both are failures of *Hormē* to align with the causal structure of reality. The pursuit of accurate models and the pursuit of effective action are one and the same.

This chapter defends that convergence. It grounds the identity of truth and goodness in the operational architecture that has been built up across the preceding propositions: a *Kriterion* that compares, and an *Organon* that acts, coupled inseparably in a single loop. A model that misjudges the causal structure of the environment generates actions that fail to maintain the boundary; a model that tracks accurately generates actions that sustain it. The comparison and the action are two moments of the same process, and their shared outcome—persistence or dissolution—is both an epistemic verdict on the model and a practical verdict on the deed.

13.1 The Separation and Its Cost

Historically, epistemology and ethics have been treated as separate branches of philosophy. Epistemology asks: what is it to know, and how can we justify our beliefs? Ethics asks: what is it to act rightly, and what makes an action good? The two inquiries rarely touch. A mathematician may possess true beliefs about prime numbers while being a moral monster; a saint may possess a pure heart while holding false beliefs about cosmology. The apparent independence of truth and goodness is one of the deep intuitions of modern thought.

This separation has a cost. If truth and goodness are independent, then there is no reason, in principle, why a perfectly accurate understanding of the world should conduce to flourishing, nor why a miserable falsehood should lead to ruin. One can imagine, with the cynic, that the lie is more adaptive than the truth, and that the virtuous are precisely those who refuse to face reality. The separation thus opens an unbridgeable chasm between the descriptive sciences—which pursue truth—and the normative disciplines—which pursue the good. Hume's Guillotine formalizes that chasm: no amount of factual knowledge can, by itself, tell us what we ought to do.[1]

This derivation challenges this entire picture. It does so not by proposing a new moral theory on top of the old epistemology, but by excavating the foundations of both and finding that they rest on a single operation.

13.2 The Operational Convergence of Model and Value

Recall the architecture of the active engine. The *Kriterion* (Proposition 5) is the physical locus where the system's internal state covaries with the external world in a manner relevant to boundary maintenance. The operational model (Proposition 6) is the structured coupling between sensor states and *Organon* outputs. This model is the system's *truth-tracker*: it is the mechanism by which the system registers what the environment is actually like.

[1]David Hume, *A Treatise of Human Nature* (John Noon, 1739), Book III, Part I, Sec. I

Now recall the source of value (Proposition 10). The good for the system is whatever fulfills its *Hormē*. But *Hormē* is fulfilled only if the system's actions succeed in biasing the causal flow toward boundary maintenance. And those actions succeed only if they are appropriate to the actual state of the world. If the model misrepresents the environment—if the bacterium's chemoreceptor binds a poison as though it were a nutrient—the ensuing action will frustrate *Hormē*. The result is not merely an epistemic mistake; it is a practical failure, a step toward dissolution.

Thus, the model is simultaneously the seat of truth and the engine of goodness. An accurate model is one whose discriminations correspond to the causal structure of the world; a good action is one that fulfills *Hormē*. But the action is generated by the model. If the model is accurate, actions tend to be good. If the model is inaccurate, actions tend to be bad. The same physical variable—the fidelity of the *Kriterion*'s coupling to the environment—determines both epistemic success and practical success.

This is not a claim that every true belief is useful or that every useful belief is true in some abstract sense. It is a claim about the *structure of agency*: the *Kriterion* that guides action must, on pain of dissolution, track the causal regularities that actually impinge on the boundary. There is no separate channel for "truth" to enter the system, and no separate channel for "value" to guide action. There is only the *Kriterion* loop, and it either tracks reality well enough to keep the system alive, or it does not.

13.3 The Collapse of the Is–Ought Gap

Hume's Guillotine asserts that one cannot deduce an "ought" from an "is." The argument is that no set of descriptive premises—statements about how the world *is*—logically entails a normative conclusion—a statement about what one *ought* to do. The gap seems absolute as long as one considers inert facts about rocks, stars, and chemical reactions.

A bounded, far-from-equilibrium agent is not a static "is" like a rock. Its "is" is an ongoing process that includes a constitutive imperative: it *must* strive to persist, or it ceases to be. The fact of being an agent is already a fact that contains a vector. To describe the agent accurately is to describe a system that is permanently in the business of pursuing certain outcomes and avoiding others. The "ought" is not a new logical operator applied to the description; it is an implication of the description's content.

Specifically, from the statement "X is an agent" (meaning: X is a bounded, far-from-equilibrium system whose *Hormē* drives a continuous comparison-action loop for boundary maintenance), we can derive the statement "X must perform actions that, on net, maintain its boundary, or X will dissolve." This "must" is a constitutive necessity, not a moral command. But as Proposition 10 argued, this constitutive necessity is the only non-mystical anchor for the concept "good." Therefore, the descriptive fact "X is an agent" directly entails the normative fact "certain outcomes are good for X (those that fulfill *Hormē*) and others are bad (those that frustrate it)." There is no gap because the description of an agent already includes the imperative to persist.

The Is–Ought gap is thus dissolved not by a trick of logic but by a re-examination of what an "is" can be. If the only "is" one considers is the static arrangement of inert matter, the gap is real. But living agents are not static arrangements of inert matter. They are processes organized around the maintenance of a boundary, and that organization *is* the ought, physically enacted. The derivation does not derive an "ought" from a bare "is" in the Humean sense; it shows that the very category of "bare is" does not apply to living systems. To exist as an agent is to be already committed.

13.4 The Functional Identity of Error and Vice

If truth and goodness are operationally convergent, then epistemic error and ethical vice are functionally identical. Both are states in which the system's internal loop fails to align with the causal order, resulting in boundary degradation.

Consider a simple case. A bacterium that has mutated such that its chemoreceptor no longer distinguishes glucose from acid. The bacterium's model of its chemical environment is now inaccurate. It swims randomly, consuming energy without bias toward nutrients. This is an epistemic failure—the bacterium's internal state does not track the truth about its surroundings. It is also, simultaneously, a practical failure—the bacterium's *Hormē* is frustrated; it will starve or poison itself. The same malfunction is both a false map and a bad outcome. There is no need to add a separate moral evaluation; the physical dissolution of the boundary *is* the negative valuation.

Scale up. A human being who harbors a distorted model of social reality—perhaps a paranoid conviction that all strangers are threats—will act on that model. The actions will be unnecessarily hostile, alienating potential allies and provoking retaliation. The result is a practical worsening of the person's condition, a frustration of *Hormē* at the social level. Again, the error and the harm are not separate events; the error *is* the generator of the harm. The derivation thus provides a unified diagnostic: wherever one finds persistent suffering, failure, or dissolution, one is likely to find a model that does not fit the causal grain of the world. And wherever one finds a model that fits precisely, one expects flourishing, not because the universe is benevolent but because accurate coupling is the only viable strategy for persistence.

This identity does not imply that every truth is immediately useful or that every falsehood immediately kills. As Proposition 4 noted, a system needs only a net positive bias. An agent can harbor many false beliefs that are neutral in their effect on persistence—false beliefs about remote galaxies, for example—and suffer no immediate boundary damage. But false beliefs that enter the comparison-action loop at the points where persistence is at stake will, on average, degrade the boundary. And over sufficient time, the bias must be toward truth, or the lineage ends.

13.5 The Pursuit as a Single Activity

The practical upshot is that the pursuit of knowledge and the pursuit of the good are not two separate human projects. They are the same project, viewed from different angles.

The scientist who seeks to understand the mechanisms of a disease is performing an epistemic task: build a model that accurately tracks the causal structure of the pathogen and the immune system. The physician who administers a treatment based on that model is performing an ethical task: act to restore the patient's *Hormē*-fulfillment. But the two tasks are inseparably joined. The model without the action is inert; the action without the model is malpractice. The entire enterprise, from laboratory to bedside, is a single extended comparison-action loop, driven by the constitutive imperative to maintain the boundaries of the human organisms involved.

Similarly, the individual who cultivates self-knowledge—who seeks to understand his own patterns of thought, emotion, and behavior—is simultaneously pursuing truth (about himself) and goodness (for himself). The accurate self-model enables more effective navigation of the social and physical world. The inaccurate self-model leads to repeated failure, suffering, and, in the limit, dissolution.

Why, then, do these pursuits so often feel decoupled? Why does the pursuit of abstract truth seem entirely divorced from physical survival? The illusion of separation is a byproduct of our own success. Because our lower-order loops—our autonomic and metabolic systems—are so ruthlessly efficient at maintaining our physical baseline, we operate with a massive boundary margin. This metabolic luxury buys the *Nous* the free time to model abstract realities, and the physical buffer to endure higher-order epistemic failures without suffering immediate dissolution. We can afford to hold false beliefs about cosmology or abstract philosophy precisely because our lower drives are so perfectly aligned with the causal grain of the world. But at the points where the boundary is actually at stake, the gap vanishes.

Socrates' dictum that the unexamined life is not worth living is a statement of *Kybernēsis* fact: a life without self-examination is steered by a model that has not been calibrated to reality, and such a life will crash against the rocks of *Dikē*—the impersonal causal consequences of error.[2]

13.6 Clarifications and Forward References

The identity of truth and goodness is the most counterintuitive claim in the derivation, and it attracts objections from evolutionary psychology, game theory, and moral experience. We pause here only to remove immediate obstacles; the full steel-manned treatment of these challenges belongs to Part VI (see Chapter 28: The Third Wall: Objections to *Hormē* and Value; and the chapter on asymmetric risk, Chapter 29: The Fourth Wall: Framework-Level Objections).

Fitness-beats-truth. The claim that evolution favors "useful fictions" over accurate maps rests on a narrow, photographic notion of truth. The framework's concept is causal truth: accurate tracking of the variables

[2]Plato, *Five Dialogues: Euthyphro, Apology, Crito, Meno, Phaedo,* trans. G. M. A. Grube (Hackett Publishing Company, 1981), 38a

that impinge on the boundary, weighted by the stakes. The "paranoid" prey animal's hair-trigger is a compressed but causally accurate map of asymmetric risk, not a falsehood. The full argument is developed in Chapter 29 and defended against objections in Part VI.

The successful psychopath. A ruthless agent who flourishes by exploiting others appears to decouple goodness from truth, but their success is parasitic on a social environment of trusting agents and is frequency-dependent. The deduction does not reduce ethics to individual persistence alone; it provides the tools to analyze the pathology as a failure of mapping at the level of the nested social boundary.

Martyrdom. Dying for a true cause can be a fulfilment of nested *Hormē* at a higher-order boundary; dying for a false or futile cause is a tragedy of epistemic failure. The derivation distinguishes the two by evaluating the accuracy of the martyr's model, not by imposing a blanket rule.

Addiction and akrasia. The apparent decoupling of true belief and bad action arises from a civil war between nested *Kriteria*—a hijacked lower-order loop versus a higher-order model that knows the danger. The truth-goodness isomorphism holds at the level of the integrated system: when the lower model dominates, the overall model is operationally false, and the boundary degrades.

Abstract pursuits. Pure mathematics, art, and philosophy may appear disconnected from survival, but they are funded by the same basal *Hormē* and serve higher-order boundaries such as intellectual legacies and cultural persistence. The system-level claim does not require every token behavior to be optimally survival-directed.

13.7 The Keystone in Place

With Proposition 11, the derivation reaches its conclusion. Life is the continuous operation of Hormē—the *Kybernēsis* of comparison and action that maintains a boundary against disruption. Agency is the same loop viewed from outside. Value is the objective relationship between the system's states and the fulfillment of that loop. And truth is the loop's accurate tracking of the causal structure that determines its success. The four concepts—life, agency, value, truth—are not separate phenomena that happen to be correlated. They are four aspects of a single architecture, logically inseparable once the boundary is posited in a disruptive world.

The Is–Ought gap collapses not because the framework has performed an illicit deduction, but because it has revealed that the "is" of a living agent already contains its own "ought." The *Kriterion* is the exact physical locus of this collapse: its material architecture is the "is," while its constitutive reference state and evaluative action enact the "ought." To be a bounded, far-from-equilibrium system is to be already committed, already striving, already in the business of distinguishing good from bad. Epistemology and ethics are not two disciplines but one, viewed from the inside and outside of the same boundary.

To assert that value is objective, however, is not to suggest that its calculation is simple. The derivation establishes the fundamental logic of the loop, but in complex agents, this architecture is profoundly nested. Molecules are subsumed in cells, cells in organs, organs in organisms, and in conscious agents, physical boundaries are nested beneath high-order conceptual and social identities. This nested architecture inevitably breeds internal friction: the *Hormē* of a lower-order component may fiercely contradict the imperative of the highest-order whole. Navigating these competing *Kriteria* is an exercise of staggering complexity. The framework does not offer a frictionless arithmetic for moral or epistemic dilemmas. What it offers is an objective coordinate system. It ensures that when we debate the "good" of a profoundly complex action, we are navigating a real, structural landscape of nested persistence, not merely trading subjective illusions.

The derivation now stands complete. Part III will deepen the mechanical picture by unpacking the *Kybernēsis* in physical detail, distinguishing teleonomy from teleology, and showing that Darwinian evolution is not an independent theory but the diachronic expression of *Hormē* across generations. But the logical spine of the book is laid. The rest is elaboration, formalization, and confrontation with the history of thought.

Part III

Mechanical Foundations

Chapter 14

The *Kybernēsis* in Physical Terms

Propositions 5 through 8 laid out the logical architecture of the comparison-action loop: a sensor that registers a current state, a reference that defines the boundary-maintaining ideal, a *Kriterion* that measures the discrepancy, and an *Organon* that acts to reduce it—all running continuously. This is the engine of *Hormē*. But "comparison" and "action" can sound abstract. To be fully convincing, the loop must be shown to be physically instantiable, and instantiated, at every scale of life. This chapter makes the loop tangible. It traces the same *Kybernēsis* architecture from the molecular machinery of a bacterium to the neural circuitry of a human brain, and shows that the logic is identical even as the material substrate changes beyond recognition.

14.0.1 The Minimal *Kriterion* in Practice: Bacterial Chemotaxis

The best-studied example of a minimal biological *Kriterion* is the chemotaxis system of *Escherichia coli.*[1]

- **Sensors.** Transmembrane receptors on the cell surface bind attractant or repellent molecules directly from the environment.
- **Reference.** Inside the cell, a signaling protein cycles continuously between two states. In the absence of attractant, the rates of activation and deactivation settle into a steady baseline—the system's "expected" condition.
- ***Kriterion.*** When an attractant binds to a receptor, the receptor changes shape, which shifts the balance of the two states away from baseline. A repellent shifts it in the opposite direction. The deviation of the current signal from the baseline is not calculated by a brain; it is physically embodied in the relative concentration of the two states.

[1] Howard C. Berg, *E. Coli in Motion* (Springer, 2004).

- **Organon.** The signaling state diffuses to the flagellar motor and biases its rotation: low deviation (favorable conditions) promotes smooth swimming; high deviation (unfavorable conditions) promotes tumbling, which randomises the cell's orientation.
- **Loop closure.** Altered swimming moves the cell to a new chemical environment, changing receptor occupancy and feeding back to the signal. The cycle repeats without pause, without a controller, without consciousness.

This is a physical instantiation of Propositions 5–8: sensor, reference, *Kriterion*, *Organon*, continuous cycle. No consciousness, no nervous system, no deliberation. Yet the effect is a clear bias up nutrient gradients and away from toxins—directed work that maintains the boundary against dissolution.

14.1 The Homeostatic Loop: Body Temperature Regulation

Scaling up, the *Kriterion* principle is preserved but elaborated. Human thermoregulation:

- **Sensors:** Thermoreceptors in the skin (peripheral) and in the preoptic anterior hypothalamus (central) provide continuous input about ambient and core temperature. [2]
- **Reference:** The hypothalamic set-point, approximately 37°C (98°F), instantiated by the temperature sensitivity of specific neurons. Their firing rate at 37°C constitutes the reference signal.
- ***Kriterion*:** The hypothalamus integrates afferent temperature signals and compares the integrated value to the set-point. The discrepancy is a neural error signal.
- **Organa:** If core temperature falls below set-point, heat-conservation *Organa* activate: cutaneous vasoconstriction, shivering, behavioral drives (seeking shelter). If core temperature rises above set-point, heat-loss *Organa* activate: vasodilation, sweating, behavioral drives (seeking shade).
- **Loop closure:** The *Organon* actions alter body temperature, which is then re-sensed, closing the loop.

[2] Walter B. Cannon, *The Wisdom of the Body* (W. W. Norton & Company, 1932).

The *Kriterion* is a distributed neural circuit, not a single protein, but the functional logic is identical. The set-point is genetically encoded and maintained by the circuit's architecture. The loop runs continuously; even during sleep, temperature is regulated.

14.1.1 The Cognitive *Kriterion*: A Human Decision

The *Kriterion* architecture does not change when it moves from chemistry to consciousness; it simply gains richer models and longer timescales.

- **Sensors.** You perceive a state of affairs—say, the room is growing dark as evening falls.
- **Reference.** An implicit standard of comfort or productivity: the room should be well-lit for reading.
- ***Kriterion.*** A discrepancy appears between the perceived dimness and the reference illumination.
- **Organon.** You stand up and switch on the lamp.
- **Loop closure.** The room brightens; perception updates; the discrepancy vanishes. The loop closes, and you return to reading.

Here, the *Kriterion* is no longer a chemical pathway or a neural circuit but a conscious deliberation. Yet the logical architecture—sensor, reference, comparison, *Organon*, loop closure—remains exactly the same. The difference is not one of kind but of richness: the model has expanded to include abstractions like comfort and productivity, and the action is guided by a representation of a desired future state rather than a direct chemical gradient. The loop still runs to closure, and the entire architecture remains that of a bounded agent navigating reality.

14.2 Scales of the *Kriterion*

The three examples illustrate a general principle: the *Kybernēsis* is scale-invariant in its logical structure. It can be instantiated in:

- **Molecular scale:** A single protein or signaling cascade (chemoreceptor, heat-shock response).
- **Cellular scale:** The integrated signaling network of a single cell (bacterium, neuron, immune cell).
- **Organ scale:** A homeostatic circuit (hypothalamus-pituitary-adrenal axis).

- **Organism scale:** The integrated behavior of the whole creature (predator stalking prey, human navigating a career).
- **Supra-organism scale:** Social systems (markets correcting prices, democratic systems correcting policy). These are higher-order boundaries with their own *Kriteria*, though whether their *Hormē* is intrinsic or derived is a separate question.

The substrate changes—proteins, neurons, institutions—but the functional logic remains: sense, compare, act, repeat. This invariance justifies the framework's substrate-neutrality. Life is not a specific chemistry; it is a specific feedback architecture.

14.3 Cybernetics versus *Kybernēsis*

When modern information theory formalized the mathematics of feedback, it adopted the term cybernetics—derived from the Greek *Kybernētēs*, meaning "steersman." But standard cybernetics abstracted the steering away from the ship. It modeled the logic of sensors, comparators, and actuators as pure, disembodied mathematics. In a standard cybernetic system—like a thermostat or a target-seeking missile—the control loop is entirely indifferent to its own material substrate. The missile executes its feedback loop even as the execution guarantees its own physical destruction. The math operates, but it possesses no constitutive anchor.

This is why this framework relies on the ancient, un-abstracted concept of *Kybernēsis*. *Kybernēsis* is not merely a mathematical control loop; it is a control loop that is ontologically bound to the boundary that houses it. In a living system, the *Kriterion* does not compare floating variables; it compares the physical state of the organism against the threshold of its own dissolution. The system is characterized by self-referential closure: the loop maintains the boundary, and the boundary protects the loop.

To conflate the two is to mistake the map for the territory. Modern cybernetics describes the how of a feedback mechanism, which can be applied equally to living agents and dead machines. *Kybernēsis* describes the what and the why—a physical loop whose very reference state emerges from the thermodynamic imperative to persist. A system only crosses the threshold into life when its cybernetic mathematics are fully subordinated to the constitutive maintenance of its own boundaries.

14.4 Looking Ahead

The *Kybernēsis* is now physically concrete. We have seen it in molecular, neural, and cognitive hardware. The next question is: what does it mean for such a loop to be "directed" without invoking purpose? That is the problem of teleonomy, the subject of Chapter 15.

Chapter 15

Teleonomy Without Teleology

The derivation has repeatedly used the language of direction. Propositions 4 and 5 spoke of work "directed" toward boundary maintenance, of comparisons that "guide" action, of a loop that "steers" the causal flow toward persistence. This language is natural, but it raises a serious philosophical question. In what sense is the work directed, if there is no mind, no intention, no future goal pulling the system forward? Are we smuggling in purpose through the back door?

This chapter answers that question by drawing a sharp distinction between two concepts that are often conflated: *teleology* and *teleonomy*. The framework requires only the latter, and the latter is entirely compatible with a world of proximate physical causation. Directed work is not work aimed at a future state; it is work shaped by a present architecture that has been stabilized by past persistence. The direction is real, but it is backward-looking, not forward-pulling.

15.1 The Two Kinds of Purpose

Teleology, in its classical sense, is explanation by final causes. The oak tree grows *in order to* provide shade; the heart beats *in order to* circulate blood; the universe unfolds *in order to* realize some cosmic end. In teleological explanation, the future outcome explains the present process. The goal is causally prior to the mechanism.

This mode of explanation has been largely banished from the natural sciences since Bacon and Galileo. It was replaced by efficient causation: every event is the necessary consequence of prior events plus the laws of nature. There is no future pulling; there is only the present pushing. The banishment was a triumph. It also left a vacuum. Organisms act *as though* they have purposes. The bacterium swims *as though* it wants to reach glucose. The heart beats *as though* it aims to circulate blood. If we refuse all teleological language, we are left describing motions without goals, and the description feels incomplete.

Teleonomy fills the vacuum. The term was coined by Colin Pittendrigh[1] and later popularized by Jacques Monod[2], who used it to describe apparent purpose arising from feedback-stabilized mechanisms. A teleonomic system is one whose behavior appears goal-directed, but whose goal-directedness is entirely explainable by the system's present physical organization and its history of differential persistence. The "goal" is not a future state exerting backward causation; it is a reference state physically encoded in the system's architecture, whose origin lies in the selective retention of past variants that happened to maintain that architecture.

The thermostat does not aim *at* a temperature; it is simply wired such that deviations from the set-point close a circuit. The bacterium does not aim *at* glucose; its chemoreceptor is simply structured such that attractant binding alters flagellar rotation. The "purpose" is a property of the mechanism, not a force in addition to it.

15.2 The Physical Basis of Teleonomy

How does a system come to have a teleonomic organization? The answer is in the derivation already completed.

Proposition 3 drew the distinction between passive conduits and active engines. An active engine is one whose internal organization biases the causal flow toward its own persistence. That bias is not a future intention; it is a present statistical tendency of the engine's energy deployment. The engine persists because its architecture yields directed work; it has that architecture because lineages that lacked it dissolved.

Proposition 5 established that directed work requires a *Kriterion*—a physical device that discriminates between persistence-relevant states. The *Kriterion* has a reference state against which current conditions are measured. This reference state defines what the system "treats as" the target condition. The treatment is not mental; it is mechanical. A protein's binding affinity, an ion gradient's equilibrium potential, a neuron's firing

[1] Colin S. Pittendrigh, "Adaptation, Natural Selection, and Behavior," in *Behavior and Evolution*, ed. Anne Roe and George Gaylord Simpson (Yale University Press, 1958), p. 394

[2] Jacques Monod, *Chance and Necessity: An Essay on the Natural Philosophy of Modern Biology* (Alfred A. Knopf, 1971).

threshold—these are reference states. They are physically present features of the system, causally responsible for its differential response to the environment. When the system acts to close the gap between the current state and the reference state, it exhibits teleonomic behavior.

Proposition 7 named the constitutive imperative *Hormē*. *Hormē* is not a desire; it is the physical fact that a far-from-equilibrium system must continuously run the *Kriterion-Organon* loop, or dissolve. The "must" is constitutive, not teleological. The system does not strive because it foresees dissolution; it strives because the alternative is immediate and physical. The future plays no role except as the inevitable terminus of a trajectory that the system's present activity deflects.

Thus, teleonomy is the shape of *Hormē* viewed from the perspective of an observer. We see the bacterium moving up the gradient, and we say it is "seeking" nutrients. But the full causal story contains only the binding of molecules, the phosphorylation of proteins, the rotation of flagella. The "seeking" is a summary of the causal structure, not an additional force.

15.3 Teleonomy and the Collapse of Is–Ought

The distinction between teleology and teleonomy is critical for understanding Proposition 11's collapse of the Is–Ought gap. If the framework were secretly teleological, it would indeed smuggle in an "ought" through the back door: the organism *ought* to persist because the universe somehow *intends* its persistence. But the framework is strictly teleonomic. The "ought" is not a cosmic command; it is a constitutive fact about the organism's own architecture. The organism is the kind of entity whose physical organization treats certain outcomes as terminal and others as preservative. That treatment is the objective basis of value. No final causes are required.

This is why the derivation can claim that value is objective without invoking a divine lawgiver or a mysterious moral property. The objectivity of value rests on the objectivity of the causal relationship between the system's organization and its persistence. That relationship is as factual as the relationship between a thermostat's set-point and the room temperature. The difference—and it is a crucial difference—is that the living system's reference state is not externally imposed but is the condition of its own continued existence. The teleonomy is intrinsic, not derived.

15.4 Clarifications and Forward References

The concepts of teleonomy and teleology invite several natural challenges: that directed work is teleology in disguise, that functional complexity implies a designer, that "teleonomy" is a semantic trick, and that the origin of teleonomy is circular. A brief clarification here removes immediate obstacles; the full steel-manned treatment of these objections belongs to Part VI (see Chapter 29: The Fourth Wall: Framework-Level Objections).

The teleology charge. The "must" of *Hormē* is constitutive, not moral. "Striving" labels the continuous operation of the *Kriterion-Organon* loop; it implies no forward-looking purpose. The distinction between teleology (future goals shaping the present) and teleonomy (present architecture shaped by past selection) is causal, not merely verbal.

The designer objection. The watchmaker argument treats organisms as tools with externally imposed reference states. The derivation shows that reference states arise constitutively from the system's own boundary condition, and that natural selection—the diachronic expression of *Hormē*—fully accounts for the emergence of functional complexity. The argument collapses once the origin of apparent design is understood.

The semantic objection. The framework's functional language is shorthand for fully specifiable proximate causal chains, exactly analogous to "attractors" and "equilibria" in physics. The difference between teleology and teleonomy is not a relabeling but a fundamental shift in the causal model.

The origin of teleonomy. How the first *Kriterion* emerged is an empirical question about abiogenesis, not a logical flaw. Self-organizing systems can exhibit primitive teleonomic behavior through physical attractors, without prior selection. The derivation requires only that a pathway is possible, not that we know the exact historical route.

15.5 Direction Without Mystery

The framework can now legitimately use the language of direction, striving, and purpose, because it has defined these terms in strictly teleonomic fashion. The direction is in the architecture, not in the future. The striving is the loop running, not a ghostly yearning. The purpose is the functional consequence of the architecture, not a cosmic plan.

With teleonomy secured, the stage is set for the full development of Darwinian evolution as the diachronic expression of *Hormē.* That is the subject of Chapter 16. The logical foundations are complete; what remains is to show how the synchronic imperative of boundary maintenance, when stretched across time, generates the tree of life.

Chapter 16

Darwinian Evolution as a Diachronic Consequence

The derivation of life, agency, and value in Propositions 1 through 11 painted a synchronic picture: a bounded system, here and now, maintaining its boundary through the continuous operation of *Hormē*. But every individual agent is finite. Sunlight fades, resources dwindle, damage accumulates beyond repair. No matter how perfectly the loop runs, the physical vehicle that houses it will eventually fail. If the phenomenon of life were confined to individual agents, it would be a brief, local flicker.

Yet life on Earth has not flickered out. It has persisted for billions of years, across countless individual deaths, spreading into every conceivable niche. To explain this, we do not need a new principle. We need only to extend the logic of *Hormē* across generations. When we do, Darwinian evolution emerges not as an independent theory grafted onto biology, but as the diachronic expression of the same constitutive imperative that operates in every living moment.

Theorem (Darwinian Evolution as a Deductive Consequence). Given the finite capacity of individual agents to resist disruption (N), the mathematical necessity of replication to distribute boundary risk over deep time, the logical requirement of variation in a changing world, and the constitutive imperative of *Hormē*, natural selection—the differential persistence of lineages whose comparison-action loops more effectively bias outcomes toward boundary maintenance—follows with deductive necessity.

This chapter unpacks that deduction. It does not depend on any particular mechanism of inheritance. Genetic mutation, epigenetic modification, cultural transmission—whatever faithfully reproduces the organization of the loop across time—will serve. The framework is agnostic about the material substrate of heredity. What it proves is that, given any substrate that permits replication with variation, the logic of *Hormē* guarantees the emergence of adaptive complexity.

16.1 From Finite Agents to Lineages

Proposition 7 established that a bounded, far-from-equilibrium system *must* strive to persist. Proposition 8 established that this striving is continuous. But nothing in those propositions guaranteed that this striving could succeed indefinitely within a single physical vehicle.

A singular active engine can successfully maintain its boundary in the short term by generating directed work (Φ) to offset environmental disruption (N). However, when extended across deep time, the calculus shifts from immediate biology to cumulative probability. Over a timeline of millions or billions of years, the probability of a catastrophic N-spike—an environmental disruption (whether thermal, kinetic, chemical, or systemic) whose magnitude exceeds the maximum possible Φ output of the organism—approaches absolute certainty. A singular, contiguous boundary is mathematically guaranteed to dissolve. Death is not a violation of *Hormē*; it is simply the inevitable statistical intersection where undirected disruption finally outpaces the directed work of maintenance.

The intuition that a singular system cannot survive deep time is an architectural reality observed across multiple disciplines. Biologists like Richard Dawkins identified the precursor to natural selection as the "survival of the stable," recognizing that physical carriers eventually break down and require replication to carry information forward.[1] Mathematicians like John von Neumann proved in the theory of self-reproducing automata that any complex, singular machine will eventually degrade due to entropy unless it possesses a universal constructor to replicate itself.[2] Statisticians using the Galton-Watson branching process demonstrated

[1]See Richard Dawkins, *The Selfish Gene* (Oxford University Press, 1976), 12–15. Dawkins introduces "survival of the stable" as the universal precursor to biological evolution, noting that complex material organizations will inevitably degrade unless they act as templates for their own replication.

[2]See John von Neumann, *Theory of Self-Reproducing Automata*, ed. Arthur W. Burks (University of Illinois Press, 1966), 74–87. Von Neumann mathematically demonstrates that mechanical or biological systems below a specific threshold of complexity inherently degrade; diachronic persistence requires an explicit mechanism of self-reproduction.

that if a system's reproduction rate does not strictly exceed one, its probability of ultimate extinction is 100%.[3]

These disciplines point toward a unified causal logic: to survive deep time, a system must distribute its boundary risk. It must spawn parallel boundaries.

We can therefore apply a biological inversion of the cosmological Anthropic Principle[4] to form a Cartesian proof of evolutionary necessity. The deduction proceeds as follows:

- We observe a continuous lineage of life existing today.
- A singular physical boundary is mathematically guaranteed to fail against the cumulative probability of catastrophic N-spikes over deep time.
- Therefore, the existence of life today mathematically necessitates that the original boundaries achieved the capacity to spawn parallel redundancies.

Reproduction is not a separate "function" of life that somehow got added to the metabolic core. It is the fundamental structural requirement for an active engine facing an inevitable N-spike. Life did not merely stumble upon reproduction; life *is* the subset of boundaries that successfully achieved replication, because anything less was systematically zeroed out by the disruptions of deep time. An agent that reproduces is an agent whose *Hormē* acts beyond the boundary of its own body.

16.2 The Logical Necessity of Variation

Could a lineage, having achieved replication, produce perfect copies of itself in perpetuity? In a perfectly static environment, the answer might be yes—an identical copy would function exactly as well as the original, and

[3]The Galton-Watson process proves that any branching demographic or biological process with an expected offspring rate of $\mu \leq 1$ faces certain extinction as time approaches infinity. See H. W. Watson and Francis Galton, "On the Probability of the Extinction of Families," *Journal of the Anthropological Institute of Great Britain and Ireland* 4 (1875): 138–44, 138–44.

[4]See Brandon Carter, "Large Number Coincidences and the Anthropic Principle in Cosmology," in *Confrontation of Cosmological Theories with Observational Data*, ed. M. S. Longair (Reidel, 1974), 291–98. The NPN framework uniquely applies this principle diachronically to biological boundary persistence rather than synchronically to cosmological fine-tuning.

the lineage could persist unchanged indefinitely. But the environment is not static. Proposition 3 already acknowledged that causal flow is regular but not frozen: the world contains variation in temperature, resource availability, chemical composition, and the activities of other active engines. Even if these fluctuations are small, they accumulate over deep time. Eventually, the environment will shift into a configuration that is incompatible with the static pattern's capacity for *Hormē*. A lineage that can only produce perfect copies has, by definition, no capacity to adapt. When the environment changes beyond the tolerance of its fixed architecture, the entire lineage dissolves.

Thus, variation is not merely an empirical accident of imperfect copying machinery; it is a *logical prerequisite* for diachronic persistence in a universe where environments are not guaranteed to remain constant. A lineage that lacks the capacity to generate structural variants is a lineage that has placed a bet on eternal stasis. The odds of that bet paying off across geological timescales are vanishingly small. Conversely, a lineage that can produce variants—copies that differ from the original in their comparison-action architecture—has, built into its very mode of replication, the capacity to stumble upon configurations that are better suited to the new conditions. Even if variation is blind, it provides the raw material for differential persistence. Without variation, there is no raw material; adaptation is impossible; extinction is certain.

Furthermore, the necessity of variation and differential replication is the only physical answer to the problem of adaptive complexity. A highly functioning Kriterion-Organon loop—whether a bacterial chemotaxis network or a vertebrate visual system—represents a state of extreme statistical improbability. A precise, interlocking "modular fit" with the environment cannot assemble ex nihilo through pure thermodynamic accident. To posit that such ordered complexity could spontaneously arise in a single, non-replicating organism without an iterative learning process is to abandon physics for theology. Reproduction with variation provides the only non-miraculous, algorithmic mechanism capable of traversing the vast probability space required to build complex active engines. It is the physical engine that converts random fluctuations into functional architecture over deep time.

Note that this argument does not depend on any empirical claim about mutation rates, polymerase fidelity, or the thermodynamics of copying. It is an *a priori* argument from the nature of a changing world and the requirement of persistence. Even if a replication mechanism could be perfectly faithful—zero errors for eternity—a lineage that employed it would simply be out-persisted, in the long run, by any lineage that could produce variation.

16.2.1 The Statistical Wall of *Ex Nihilo* Complexity

The necessity of variation and differential replication provides the only physical answer to the problem of adaptive complexity. A highly functioning *Kriterion-Organon* loop represents a state of extreme statistical improbability.

To quantify this impossibility, we need only look at the sheer statistical mass of biological organization. Biophysicist Harold Morowitz calculated the probability of a single, simple bacterium assembling its covalent bonds and structural sequence by pure chance in a state of thermodynamic equilibrium. He arrived at a probability so vanishingly small—roughly 1 in $10^{100,000,000,000}$—that it renders spontaneous assembly functionally impossible in this universe.[5] Similarly, astrophysicist Fred Hoyle famously calculated the odds of merely the basic enzymes required for life forming by random shuffling, arriving at 1 in $10^{40,000}$—a statistical absurdity he compared to a tornado sweeping through a junkyard and accidentally assembling a Boeing 747.[6]

These monumental calculations do not refute the existence of complex life; they refute the possibility of its spontaneous generation in the absence of replication. An active engine cannot climb the probability mountain in a single leap.

---

[5]See Harold J. Morowitz, *Energy Flow in Biology: Biological Organization as a Problem in Thermal Physics* (Academic Press, 1968). Morowitz mathematically demonstrated that the spontaneous generation of even the simplest bounded living system requires traversing a probability gradient that is physically insurmountable without an iterative, energy-consuming process.

[6]See Fred Hoyle and Chandra Wickramasinghe, *Evolution from Space* (Simon; Schuster, 1981), 24. While Hoyle deployed this statistic to argue for panspermia, evolutionary biology correctly interprets it as the ultimate mathematical refutation of spontaneous generation. The "junkyard tornado" proves that complex active engines cannot be built in a single probabilistic leap.

Therefore, reproduction with variation is not merely an empirical habit of Earth-based life; it is the *only* non-miraculous, algorithmic mechanism mathematically capable of traversing the vast probability space required to build complex agents. By breaking the assembly of complexity down into tiny, iterative steps—where each successful functional configuration that survives an N-spike is copied and saved before the next variation is attempted—the system climbs the gradient. Reproduction is the physical engine that converts random fluctuations into functional architecture over deep time.

16.3 The Deduction of Natural Selection

With the logical necessity of variation in place, the deduction of natural selection is straightforward. It requires no new principles beyond those already established in Propositions 1–11.

1. **Finite lifespans.** Individual agents are finite; their physical boundaries will eventually dissolve as the cumulative toll of undirected disruption inevitably overwhelms the system's capacity for repair.
2. **Diachronic pattern persistence requires replication.** For the organizational pattern of *Hormē* to persist beyond the death of any individual vehicle, it must be replicated into new physical vehicles.
3. **Variation is a logical necessity.** In a world where the environment undergoes change, a lineage that replicates without variation will eventually encounter conditions to which it is unadapted and will dissolve. Therefore, the capacity to generate variation among replicated patterns is a prerequisite for long-term diachronic persistence.
4. **Differential persistence.** Among the variants produced by a lineage, those whose comparison-action loops bias outcomes toward persistence more effectively in the current environment will, on average, persist longer and replicate more frequently than those whose loops are less effective. This follows from Proposition 4 (directed work must bias outcomes toward persistence) and Proposition 5 (effective bias requires accurate comparison): a loop that better tracks the causal structure of the environment and deploys energy more efficiently will maintain the boundary's structural integrity more reliably.

5. **Cumulative change.** Over deep time, the distribution of organizational patterns within the population shifts toward those variants that are more effective at boundary maintenance. This is the mathematical inevitability of cumulative differential replication.

Point 5 is natural selection. It is not an independent empirical hypothesis grafted onto the framework; it is the diachronic shadow of the very architecture that defines life. The same *Kriterion-Organon* loop that keeps a bacterium swimming up a nutrient gradient today is the loop whose differential fidelity, across a billion generations, sculpts the machinery of the eye.

16.3.1 The Tautology Objection and Its Defeat

A perennial charge against natural selection is that it is circular: "survival of the fittest" merely identifies the fittest as those who survive, and then claims that they survive because they are fit. If the deduction above were guilty of this circle, it would explain nothing.

The circle is broken by observing that *Hormic effectiveness*—the property that makes one variant "fitter" than another—is defined independently of actual survival. Propositions 1–8 define what it means for a system to be architecturally capable of maintaining its boundary: its *Kriterion* must accurately discriminate persistence-relevant states (Proposition 5), its *Organon* must direct work appropriately (Propositions 4 and 6), and its loop must run continuously (Proposition 8). These are measurable, physical properties—the fidelity of sensory coupling to environmental variables, the efficiency of energy transduction, the latency of corrective action. A variant whose architecture scores higher on these operational metrics is *architecturally more effective* at boundary maintenance, regardless of whether it happens to survive a particular environmental roll of the dice.

The deduction then asserts that, statistically, such variants will out-persist others across large numbers and deep time. This is a physical prediction, not a tautology. It says: measure the operational properties of a lineage's comparison-action loop; from those measurements, compute its expected capacity to buffer against a given distribution of environmental disruptions; and compare it to another lineage. The lineage with the higher expected resilience will, with overwhelming probability, persist longer. If this prediction were falsified—if a lineage with demonstrably poorer comparison-action architecture routinely out-persisted a lineage with richer architecture in identical environments—the framework would collapse. No such falsification has ever been observed.

Thus, natural selection is not an empty truism. It is a deductive consequence of the physical requirements of boundary maintenance, tied to an operational definition of fitness that stands on its own structural feet.

16.4 Substrate Agnosticism: Why the Mechanism of Inheritance Does Not Matter

The deduction of natural selection from the logic of *Hormē* makes no commitment to any particular mechanism of inheritance. The framework does not need to specify whether variation arises from random genetic mutation, epigenetic modification, horizontal gene transfer, cultural transmission, or any other process. The only requirements are that:

- The organizational pattern (the architecture of the comparison-action loop) is replicated with some degree of fidelity across generations.
- The replication process introduces variation in that pattern.
- Variants are subject to the same constitutive imperative of boundary maintenance in a shared environment.

Every known and conceivable inheritance system satisfies these requirements. If epigenetic changes acquired during an organism's lifetime can be transmitted to offspring, that is simply an additional channel of variation, and variants with adaptive epigenetic modifications will be differentially preserved. If cultural practices can be transmitted, variants with adaptive cultural traits will be differentially preserved. The logic of differential persistence is entirely insensitive to the material substrate of the traits being selected.

This agnosticism about the specific material substrate of inheritance is a strength, not a weakness. The derivation does not merely accommodate any mechanism; it predicts that, wherever multiple inheritance mechanisms compete, the one—or the combination—that most effectively preserves and transmits *Hormic* patterns across generations will be differentially retained. Over deep time, the system converges on whatever works best, whether that turns out to be genetic mutation, epigenetic modification, cultural transmission, or some interplay of all three.

The recent resurgence of interest in epigenetic inheritance, niche construction, and developmental plasticity does not threaten the deduction; it is exactly what the logic of diachronic *Hormē* would lead us to expect. The details are for empirical biology to fill in; the invariant logical structure—differential persistence of the most effective *Hormic* patterns under disruption—is already complete.

16.5 The *Logos*, *Dikē*, and the Imperative of Adaptation

The deduction of natural selection from *Hormē* (Section 16.3) demonstrated that differential persistence follows mathematically from finite lifespans, replication with variation, and the constitutive imperative. But it did not yet answer a deeper question: why must the *Kriterion-Organon* loop track reality at all? Why does a failure to track the causal structure of the world result in dissolution, whether for the individual or the lineage?

The answer lies in two concepts that the derivation has implicitly relied on from the beginning, and that the larger NPN framework makes explicit: *Logos* and *Dikē*.

The *Logos* is the causal structure of reality. It is the inherent, discoverable order that governs all behavior within the *Archē*—the laws, regularities, and constraints that make minimal determinism (Chapter 19: The Ontological Mandate of Minimal Determinism) possible. The *Logos* is not a conscious mind or a divine planner; it is simply the way the world is organized, the grain along which events unfold. A bacterium swimming up a chemical gradient is navigating the *Logos* at the scale of molecular diffusion. A predator tracking prey is navigating the *Logos* at the scale of biomechanics and ecology. A lineage adapting over millennia is navigating the *Logos* at the scale of environmental change and competitive dynamics. In every case, the *Logos* is what the organism's *Kriterion* must model—accurately enough—if its actions are to bias outcomes toward persistence.

***Dikē* is the enforcement of the *Logos*.** *Dikē* is not a moral punishment; it is the impersonal, causal consequence of misalignment. If an organism's *Kriterion* misjudges the gradient—if it treats poison as nutrient, or a predator as a shadow—the boundary degrades. *Dikē* is the impersonal consequence that follows. If a lineage's architecture cannot track a shifting environment, *Dikē* is the extinction that removes it from the record.

There is no cosmic judge, no wrathful arbiter—only the impartial operation of the *Logos* on whatever configurations of matter happen to exist. *Dikē* is what happens when a false map meets the real territory, and the territory always wins.

The imperative to track reality, then, is not a separate requirement added to the deduction. It is built into the relationship between *Hormē* and the world it navigates. *Hormē* is the striving; the *Logos* is the structure of what is striven within; *Dikē* is the cost of erring. Organisms do not merely benefit from accurate models—they are *required* to have them, on pain of dissolution. The *Kriterion* that fails to correlate its discriminations with the causal structure of the environment is a *Kriterion* that will soon belong to a dead organism. The lineage that fails to track the shifting *Logos* across generations is a lineage that will leave no descendants.

This is why variation and adaptation are not optional strategies. The world changes—not because it is malevolent, but because the *Logos* includes flux. Climates shift, resources dwindle, competitors evolve, new pathogens emerge. A static architecture is a bet that the environment will remain constant, and across deep time, that bet is always lost. The capacity to generate variation—to explore new configurations of the *Kriterion-Organon* loop—is a prerequisite for diachronic persistence precisely because the *Logos* is dynamic, and *Dikē* is patient but absolute.

16.6 Evolution as Deep-Time Navigation

Evolution is not a blind watchmaker stumbling upon adaptations by accident. It is a deep-time navigational strategy, structurally identical to the moment-by-moment *Kybernēsis* loop that operates within each individual organism.

At the synchronic scale, an organism navigates its environment by continuously comparing its current state to a reference state, acting to reduce the discrepancy, sensing the result, and repeating. Over a lifetime, this loop allows the organism to track and exploit the causal regularities of its world.

At the diachronic scale, a lineage navigates its environment by continuously varying its own architecture, having those variants tested against the environment, and differentially retaining the patterns that succeed.

Over evolutionary time, this loop allows the lineage to track and exploit the causal regularities of a changing world.

The two loops are nested instances of the same logic. The individual's *Kybernēsis* operates over a lifetime; the lineage's evolutionary loop operates over generations. The individual's model is encoded in its neural or molecular *Kriteria*; the lineage's model is encoded in the anatomy, physiology, and behavior of its organisms. In both cases, the imperative is the same: persist by aligning internal organization with the causal structure of reality. In both cases, success is truth, and failure is dissolution.

This is why the truth-goodness isomorphism (Proposition 11) holds diachronically as well as synchronically. A lineage that systematically fails to track the causal structure of its environment will go extinct. A lineage that persists for millions of years is a lineage that has, across its evolutionary history, accumulated a portfolio of operational models that are accurate enough to sustain the boundary. The fossil record is a graveyard of false maps; the extant biosphere is a library of truths, written in the language of anatomy, physiology, and behavior.

16.7 Clarifications and Forward References

The deduction of natural selection from *Hormē* invites several challenges: that replication is an empirical premise rather than a derived necessity, that the origin of life remains unexplained, that Lamarckian inheritance undermines the claim, and that neutral drift is incompatible with selection's diachronic role. A brief clarification here removes immediate obstacles; the full steel-manned treatment of these objections belongs to Part VI (see Chapter 29: The Fourth Wall: Framework-Level Objections).

Replication as a structural necessity. The framework does not treat replication as a fortunate empirical accident. As demonstrated by the Anthropic deduction of deep time, replication is a mathematical prerequisite for the continued observation of life. The derivation proves that diachronic persistence of the *Hormic* pattern against the certainty of catastrophic N-spikes requires the distribution of boundary risk, and that natural selection is the mathematical consequence of that requirement once variation is present.

The origin of life. The derivation defines the minimal architecture life must possess; it does not narrate the historical path from prebiotic chemistry to the first active engine. That is an empirical question, and the framework sets the target for it.

Lamarckian inheritance. Natural selection here means differential persistence of heritable variants, irrespective of the mechanism of variation. Epigenetic, cultural, or genetic variation all feed into the same diachronic dynamic.

Neutral drift. Selection is the dominant signal over deep time; drift is the tolerated noise around it, consistent with Proposition 4's requirement of a net positive bias rather than perfect optimization.

16.8 The Diachronic Imperative

The constitutive imperative that drives a bacterium up a gradient is the same imperative that, across deep time, sculpts the vertebrate eye, the avian wing, and the human neocortex. The *Logos* that the bacterium must track with its chemoreceptors is the same *Logos* that selects the lineages whose architecture tracks it best. *Dikē* enforces both verdicts—dissolution for the individual that misreads the gradient, extinction for the lineage that cannot adapt.

Thus, the diachronic imperative is not a metaphor. Evolution is not a separate force; it is *Hormē* written across deep time, answerable to the same reality as the bacterium's lonely swim.

The derivation of the framework is now complete in both its synchronic and diachronic dimensions. Part IV will formalize the boundary integral into a rigorous equation and explore the predictive implications for epistemology, ontology, and the architecture of nested minds. But the spine of the argument—the eleven propositions plus the diachronic theorem—stands on its own, anchored in the logic of the boundary and answerable to reality.

Part IV

Formalism and Prediction

Chapter 17

The Mathematics of Boundary Maintenance

The derivation has proceeded conceptually. Propositions 1–8 identified the necessary architecture of a living system: a bounded, far-from-equilibrium entity that continuously operates a comparison-action loop to bias the causal flow toward its own persistence. Proposition 7 named that constitutive imperative *Hormē*, and Proposition 8 made its operation continuous. But "continuous operation" and "net positive bias" are qualitative terms. To give the framework full explanatory power—to make it falsifiable, to distinguish the living from the non-living with precision, and to guide the design of artificial systems—we must translate the conceptual logic into a formal, mathematical condition.

This is the boundary integral. It is not a new physical law; it is a formalization of the constitutive necessity. It distills the entire derivation into an equation that any system must satisfy if it is to persist as a distinct entity over time.

17.1 The Constitutive Condition Formalized

Recall the minimal condition for persistence. A bounded entity begins with an initial organizational integrity—a certain capacity to resist disruption. That capacity is continuously eroded by disruptive interactions (N), and continuously restored, or at least defended, by directed work (Φ). If, over any interval, the cumulative erosion outpaces the cumulative repair, the entity's boundary degrades. If the degradation reaches a critical point, the boundary collapses, and the entity ceases to exist as a distinct thing.

This condition can be expressed mathematically. Let $A(t)$ represent the active organizational state of the system at time t—a measure of how far the boundary is from dissolution. $A(t)$ is not a physical energy in the narrow thermodynamic sense; it is a functional index of the boundary's capacity to maintain itself. For a cell, $A(t)$ might be a function of membrane integrity, ion gradients, and repair-enzyme activity. For a multicellular organism, it might integrate homeostatic variables across organ systems. The precise form of $A(t)$ depends on the substrate, but the logic is universal.

Then the persistence condition is:

$$A(t) = A_0 + \int_0^t (\Phi(\tau) - N(\tau))\, d\tau,$$

where:

- A_0 is the initial organizational state at $t = 0$—the inherited boundary capacity. For a newborn organism, this is the structural endowment from its progenitor; for the first living system, it would be the minimal self-organizing configuration that kicked off the lineage.
- $\Phi(\tau)$ is the instantaneous rate of directed work at time τ—the net favorable alteration of the causal flow that tends to maintain or restore the boundary. This includes repair of damage, active transport of nutrients, evasion of threats, and any other energy deployment that biases outcomes toward persistence.
- $N(\tau)$ is the instantaneous rate of disruptive work at time τ—the net unfavorable alteration of the causal flow that tends to degrade the boundary. This includes external insults (mechanical stress, chemical attack, radiation) and internal wear (metabolic byproducts, entropic decay of structures).

The integral $\int(\Phi - N)d\tau$ is the accumulated net margin. If, over any critical interval, this integral becomes too negative—that is, if N exceeds Φ by more than the system's current $A(t)$ can absorb—the boundary loses coherence and the system dissolves.

The equation is not a discovery of a new force. It is the constitutive condition of identity over time, expressed in the language of rates and accumulations. It says nothing more, and nothing less, than: a bounded entity persists only if the work done to maintain it outruns the work done to destroy it.

17.2 The Threshold of Dissolution

Not every system that satisfies the equation for a while persists forever. There is a critical threshold, A_{min}, below which the boundary's organization is so degraded that no feasible directed work can restore it. For a cell, this might be the point at which membrane rupture exceeds the capacity of repair machinery, or at which the ion gradients collapse irreversibly. For a multicellular organism, it is the point of no return in circulatory failure or neural death. The threshold is a real physical quantity, even if its precise value depends on the specific architecture of the system.

Thus the full condition for continued existence is:

$$A(t) \geq A_{\text{min}} \quad \text{for all } t.$$

If at any time t, $A(t)$ falls below A_{min}, the system ceases to exist—even if later events would hypothetically have restored it. Death is a one-way door.

This threshold gives teeth to the concept of "net positive bias." A system with a wide margin—a large A_0 relative to typical fluctuations in N—can tolerate occasional, even large, failures of directed work. A system with a narrow margin must maintain $\Phi > N$ almost continuously. The variation in margins explains why some organisms can survive extreme environments for short periods, while others die at the slightest perturbation. It also explains why complexity is a double-edged sword: a complex system can generate enormous Φ, but its many interdependencies often create a thin A_{min} margin, making it fragile to cascading failures.

17.3 Justifying the Equation from Propositions 1–8

The boundary integral is a direct translation of the propositions into mathematical form.

- **Proposition 1 (Boundary):** The existence of a boundary defines an inside and an outside, and $A(t)$ quantifies the integrity of that boundary. $A(t) > A_{\text{min}}$ means the boundary is intact; $A(t) \leq A_{\text{min}}$ means it has dissolved.
- **Proposition 2 (Disruption):** $N(\tau)$ is the mathematical face of continuous degradation.

- **Proposition 3 (Primal Distinction):** Passive conduits have $\Phi = 0$; their $A(t)$ can only decrease or, at best, remain constant if $N = 0$. Active engines deploy $\Phi > 0$ to counteract N.
- **Proposition 4 (Directed Work):** Φ must, on net, be positive and correlated with boundary maintenance. The equation captures the net bias: the integral must trend upward or remain steady over relevant timescales, not drift downward.
- **Proposition 5 (Cognitive Minimum):** The *Kriterion* is the physical mechanism that modulates Φ in response to the state of the boundary and environment. When the *Kriterion* detects that $A(t)$ is falling or that N is rising, it triggers an increase in Φ—e.g., activating repair enzymes or fleeing a threat. Without such discrimination, Φ would be uncorrelated with $A(t)$.
- **Propositions 6–8 (Loop and Imperative):** The continuous operation of the loop ensures that Φ is not applied in discrete bursts but is adjusted in real time. The constitutive imperative is the necessity of maintaining the integral condition; "must strive" means "must keep $\int(\Phi - N)d\tau$ from violating A_{min}."

Thus the equation is not an empirical add-on. It is the precise quantitative form of the logic already derived. It allows us to move from "life requires net positive directed work" to "life requires that the initial boundary value plus the time integral of $\Phi - N$ strictly exceeds A_{min}."

17.4 The *Kybernēsis* as Integral Control

The equation also illuminates why the *Kybernēsis* must be continuous and why simple set-point regulation is often insufficient. A system that only responds when $A(t)$ approaches A_{min} risks a fatal lag—the repair might come too late. A system that continuously tracks the rate of change of $A(t)$, or even anticipates future N based on environmental cues, can keep the integral safely positive with smaller, smoother corrections. This is exactly what we observe in biological regulation: anticipatory responses (feedforward) layered on top of feedback loops.

In control theory terms, *Hormē* implements a kind of *integral control*: the accumulated error (loss of boundary integrity) must be kept near zero over time, not just corrected instantaneously. The continuous *Kriterion-Organon* loop ensures that the integral never drifts too far negative.

17.5 Clarifications and Forward References

The boundary integral, like any formalization, invites questions about its empirical status, measurability, and scope. Three challenges arise: that it is merely a restatement of the Second Law, that its variables are unmeasurable, and that many complex systems lack a sharp dissolution threshold. A brief signpost here removes immediate obstacles; the full steel-manned treatment belongs to Part VI (see Chapter 29: The Fourth Wall: Framework-Level Objections).

The thermodynamic truism charge. The integral is a constitutive condition for any organized pattern, not a special case of entropy. It requires a specific functional coupling between Φ and $A(t)$ that thermodynamics alone does not supply.

Measurability. $A(t)$, Φ, and N are operationalizable for any specific system, from cells to ecosystems, and the framework provides a template for doing so independently of survival outcomes.

Sharp versus graded thresholds. A_{min} can be a region rather than a point; the equation provides a multi-scale diagnostic for the integrity of boundaries, not a binary all-or-nothing judgment.

17.6 From Equation to Prediction

The boundary integral is not merely a restatement of already-derived truths. It yields testable predictions. For instance, it predicts that systems with high N relative to Φ can persist only transiently unless they can increase Φ or reduce N through adaptation—a pattern observed in evolutionary rescue studies, where populations facing severe environmental stress either adapt rapidly or go extinct.[1] It predicts that the complexity of a system's *Kriterion* should scale with the variability of N in its niche: predictable environments allow simpler *Kriteria*, while chaotic environments demand more sophisticated predictive models. This is precisely

[1]Graham Bell, "Evolutionary Rescue and the Limits of Adaptation," *Philosophical Transactions of the Royal Society B* 368, no. 1610 (2013): 20120080, `https://doi.org/10.1098/rstb.2012.0080`.

what comparative genomics reveals—bacteria living in stable intracellular environments possess far fewer signal-transduction genes than their free-living relatives in fluctuating soil or water.[2]

In the next chapters, we will deploy the boundary integral to illuminate specific phenomena: the epistemology of asymmetric risk, the ontological mandate of minimal determinism, and the internal conflict resolution in nested *Hormic* architectures. The equation is the quantitative thread that ties the entire framework together. It transforms the philosophy of the boundary into a tool for empirical science.

[2]Michael Y. Galperin, "A Census of Membrane-Bound and Intracellular Signal Transduction Proteins in Bacteria: Bacterial IQ, Extroverts and Introverts," *BMC Microbiology* 5 (2005): 35, `https://doi.org/10.1186/1471-2180-5-35`.

Chapter 18

The Epistemology of Asymmetric Risk

A rabbit freezes at the crack of a twig. A human feels a surge of dread at the sight of a coiled shape in the corner of the garden. A social animal hesitates before challenging a dominant rival, its pulse racing at the mere thought of confrontation. In each case, the organism is responding to a cue that might signal danger—or might be nothing at all. The rabbit's twig might be a fox or the wind. The coiled shape might be a viper or a fallen rope. The rival's glare might precede a mauling or a bluff.

These responses have long been classed by psychology as *irrational*—cognitive biases, phobias, or errors in probabilistic reasoning. The rabbit that startles at every sound is wasting energy; the human with arachnophobia is overestimating objective risk. From the perspective of a disinterested epistemic agent measuring the world for its own sake, these reactions are errors of calibration. The organism sees danger where there is none.

Hormē reverses this judgment. Once we understand that the organism's cognitive apparatus is not a detached truth-tracker but a component of a boundary-maintaining engine, the very meaning of "error" shifts. The rabbit's jumpiness, the human's dread, the social animal's caution are not failures of rationality. They are exquisitely accurate operational maps of the *causal stakes* of being wrong. This chapter explains why.

18.1 The Puzzle of Over-Reaction

Evolutionary psychology has cataloged a menagerie of seemingly irrational fears. Humans are disproportionately afraid of snakes, spiders, heights, and enclosed spaces, relative to their actual danger in the modern world. A cat leaps at the sight of a cucumber. A prey animal will flee a harmless shadow that vaguely resembles a hawk, while continuing to graze near a far more statistically lethal road. These phenomena are robust, cross-cultural, and often resistant to corrective information. Even when we know the spider is harmless, our autonomic nervous system does not get the memo.

The standard cognitive science account labels these as biases—the availability heuristic, preparedness theory, error management theory. But these accounts still frame the reaction as an *error*: the organism's internal model is false, even if the falsehood is adaptive. The map says "predator" when the territory says "shadow." The map is wrong, but being wrong kept your ancestors alive.

The map is not wrong. It is perfectly accurate about the only thing that matters to the organism: the expected impact on its persistence.

18.2 The Physics of Being Wrong

Recall the boundary integral from Chapter 17: The Mathematics of Boundary Maintenance: persistence requires that directed work outpace disruptive work over time. For an organism making a perceptual judgment, every decision to act or not act feeds into that balance.

When the organism faces an ambiguous cue—something that might be a threat—it can make two kinds of mistake. It can treat a harmless thing as dangerous: jump at a shadow, startle at a twig, avoid a harmless spider. Or it can treat a dangerous thing as harmless: ignore the viper, dismiss the stalking predator, laugh off the genuine rival. The costs of these two mistakes are not just different in degree; they are different in kind.

A false positive—fleeing when there is no danger—costs a finite, usually tiny, amount of directed work. The rabbit burns some calories, loses a few moments of grazing, and returns to baseline. Its boundary margin dips slightly and then recovers. The rabbit can afford thousands of such mistakes over a lifetime; they are a minor operating expense.

A false negative—failing to flee when there *is* danger—costs the entire boundary. The rabbit is eaten. There is no recovery, no learning, no second chance. The cost is absolute: the system dissolves.

The asymmetry is not a matter of perspective or subjective valuation. It is a physical fact about the boundary's tolerance. The boundary integral specifies exactly what these costs are, in principle measurable: the false positive cost is the decrement to Φ from the unnecessary action; the false negative cost is the entire remaining $A(t)$ crashing to zero. The ratio between them, for any lethal threat, effectively approaches infinity to one.[1]

18.3 The Calculus of Survival

Game theory has long understood that optimal decisions depend on the payoff matrix. But game theory typically treats the payoffs as abstract numbers—utilities, preferences, subjective values—without a physical anchor. The boundary integral changes this. It provides a way, in principle, to *calculate* the true payoffs from the physics of the organism's boundary maintenance. The cost of a false positive isn't a matter of how the organism "feels" about wasted energy; it's the objective reduction in $A(t)$. The cost of a false negative isn't an assigned utility; it's the total loss of the system. These numbers are grounded in the measurable rates of directed work and boundary integrity.

Given this payoff matrix, the optimal strategy is not to maximize representational accuracy—to build the most photographically faithful map of the world. It is to maximize the expected value of staying above A_{min}. And because the cost of a false negative is so astronomically larger than the cost of a false positive, the optimal strategy is to treat ambiguous threats as real *unless* the cost of the avoidance itself approaches the cost of death—which it almost never does.

If a coiled shape has a tiny chance of being a viper, and the cost of stepping on a viper is death, while the cost of avoiding a rope is a trivial detour, the smart money is on the detour. Every single time. The rabbit that runs from every rustle is not paranoid; it is executing the only strategy that has a positive expected boundary margin. The rabbit that calculates the exact probability and decides to stand its ground is making a catastrophic mathematical error: it has ignored the asymmetry of the stakes.

This doesn't require doing the math consciously. Evolution has already done it. The rabbit's neural architecture, tuned over millions of generations, embodies the solution to the optimization problem. The hair-trigger is the answer.

---

[1]Randolph M. Nesse, "The Smoke Detector Principle: Natural Selection and the Regulation of Defensive Responses," *Annals of the New York Academy of Sciences* 935, no. 1 (2001): 75–85, `https://doi.org/10.1111/j.1749-6632.2001.tb03473.x`.

18.4 The "Paranoid" Map Is the True Map

The standard critique calls the startle response a "false belief" about the world. The organism's internal model represents the shadow as a predator, but the shadow is not a predator. The representation is factually incorrect. Therefore, the organism has a false map.

The organism's map is not a value-neutral representation of object identities. It is a compressed, operational encoding of *causal relevance to boundary persistence.* The map does not say "that is a predator." It says "that ambiguous stimulus has the kind of expected impact on my boundary that requires immediate flight." That statement is not false. It is perfectly true. The shadow is, in fact, the kind of thing that could kill you if it were a predator, and given the sensory limitations and the speed required, the only boundary-preserving response is to treat it as lethal until proven otherwise.

In other words, the organism is not modeling the ontology of the object (shadow vs. predator). It is modeling the *decision-relevant risk structure* of the environment. And that model is accurate. The "paranoid" prey animal that runs from every shadow has an *epistemically virtuous* map with respect to the variables that matter. The "rational" prey animal that calculates the true probability and decides not to run is, in fact, the one with a false map—it has failed to incorporate the severity of the payoff asymmetry into its calculation.

This is a direct consequence of the truth-goodness isomorphism (Proposition 11). Truth, for an agent, is not disinterested correspondence. It is the alignment of the internal *Kriterion* with the causal structure of reality, weighted by the stakes that reality imposes on the boundary. A map that outputs "run" when presented with an ambiguous stimulus in a high-stakes environment is a true map, because it correctly identifies the action that maximizes $A(t)$. Truth is what works, and what works in an asymmetric payoff landscape is a hair-trigger.

18.5 Examples Across Scales

The principle of asymmetric risk explains a broad range of phenomena that otherwise appear as quirks of psychology or evolution.

Prepared fears. Humans rapidly acquire fears of snakes, spiders, heights, and enclosed spaces, and these fears are resistant to extinction, even in environments where such threats are minimal. The modern environment is an evolutionary anomaly; for the vast majority of hominid history, a false negative with a snake or a fall from a height was catastrophic. The *Kriterion*'s reference state was tuned over deep time to an environment where getting it wrong meant death. The organism's map remains causally accurate relative to the world it was built for.

Anxiety disorders. Clinical anxiety can be understood as a *Kriterion* whose threshold for threat detection has been set too low, or whose estimate of the cost of a false negative is permanently elevated. The anxious person's model of the world is not "false" in content—social evaluation can indeed lead to rejection—but the *gain* on the threat signal is miscalibrated for the current environment. This is a map with an incorrect prior on the severity distribution, not a wholly imaginary fear.

Superstition and ritual. Pigeons in Skinner boxes develop superstitious behaviors when rewards are delivered at random intervals.[2] The cost of performing the ritual is low—a trivial energy expenditure—and the potential cost of *not* performing it, if the association were real, could be losing access to food. In an environment where the causal structure is opaque, retaining any behavior that seems to correlate with survival-relevant outcomes is the optimal strategy, given the asymmetry of the stakes. Superstition is the cognitive shadow of adaptive prudence in an information-poor world.

Social caution. The fear of public speaking often outranks the fear of death in surveys. For a social primate, social exclusion can cascade into loss of resources, allies, and protection—a genuine threat to the boundary, albeit a slow one. The *Kriterion*'s threat-detection for social evaluation is calibrated to high stakes in an ancestral context where ostracism meant existential risk. The map is not false; it is tracking the causal stakes of a nested boundary.

[2]B. F. Skinner, "'Superstition' in the Pigeon," *Journal of Experimental Psychology* 38, no. 2 (1948): 168–72.

18.6 The Ontology of "False" Maps

The argument of this chapter raises a deeper question: in what sense can any persistent organism's map ever be called "false"? If the organism is alive, its map must be true enough to have kept it alive. The framework answers: maps can be false when they fail to track the causal structure at the points where the boundary margin is thin, and when that failure results in dissolution or suffering. But they cannot be called false merely because they fail to represent the world with photographic fidelity. Fidelity is not the goal; persistence is.

A map with a hair-trigger on predator-shaped stimuli is not a false map with a lucky error. It is a map that has extracted the causal invariant—*ambiguous quick movements can kill you*—and compressed it into a perceptual algorithm. The compression discards irrelevant detail and preserves the only feature that matters: the action that maximizes survival given the limited information. This is exactly what a good map does.

18.7 Clarifications and Forward References

The claim that a hair-trigger startle response is an *operationally true* map—because it accurately tracks causal stakes—invites several challenges. Does this not render the framework unfalsifiable, since any fear can be rationalized? Does it excuse modern maladaptive fears by appealing to ancestral environments? Does it conflate pragmatic utility with truth? These questions are flagged here only briefly; the full steel-manned treatment belongs to Part VI (see Chapter 29: The Fourth Wall: Framework-Level Objections).

Over-inclusiveness. The framework provides a precise criterion: a map is operationally true only if its net effect on $A(t)$ is positive. A phobia that causes self-destructive behavior is a false map, regardless of its ancestral calibration.

Modern mismatch. A map that was accurate in an ancestral environment but is miscalibrated for the present one is false relative to current causal structure. The framework explains why recalibration lags—the *Kriterion* is slow to update—but does not excuse the error.

Truth-deflation. The framework distinguishes propositional truth (the shadow is not a predator) from operational truth (the sensor-*Organon* coupling correctly models the asymmetric risk). The former is not denied; the latter is the relevant concept for understanding agency.

18.8 The Rationality of the Frightened Animal

The rabbit that startles at the twig, the human who sweats at the mention of spiders, the anxious social climber who sees judgment in every glance—these are not broken thinkers. They are agents whose *Kriteria* are tuned to the most fundamental fact of existence: the boundary is fragile, and the cost of ignoring a threat is absolute. Their maps are not filled with errors; they are filled with the accumulated wisdom of ancestors who, facing exactly the same asymmetric payoffs, fled at the right moments, and lived.

The epistemology of asymmetric risk is a direct application of the boundary integral. It shows how the framework turns abstract game-theoretic payoffs into physically grounded, in-principle measurable costs. And it reveals that the rabbit's leap is more than a flash of nerve—it is a proof that the universe must possess enough causal order for such leaps to work. That is the subject of the next chapter: the ontological mandate of minimal determinism.

Chapter 19

The Ontological Mandate of Minimal Determinism

The rabbit that startles at the twig and survives does not merely teach us something about the psychology of fear. It teaches us something about the structure of reality itself. The rabbit's leap could not work—could not be a rational response to an asymmetric payoff—unless the world possessed a certain minimal order. If the same twig crack sometimes preceded a fox, sometimes a falling tree, sometimes a blast of music, and sometimes nothing at all, with no stable relationship between the sound and its source, then the rabbit's startle would be a pointless spasm. It could not *bias* the outcome. It would be noise meeting noise, and the rabbit's boundary would dissolve regardless of what it did.

That the rabbit has survived, that any active engine exists at all, is therefore a proof. It proves that the universe is not pure chaos. The causal flow, upon which the engine depends to redirect events toward its own persistence, must possess regularities that can be exploited. This requirement is not a physical law among others. It is an ontological precondition for the very possibility of agency.

19.1 The Problem of Pure Randomness

Proposition 3 introduced the concept of an active engine: a bounded system that harnesses energy to redirect the causal flow toward its own persistence. But what if the causal flow cannot be redirected? What if the world is such that any action, any expenditure of energy, produces outcomes that are entirely uncorrelated with the action's character—just random noise drawn from a universe with no lawful structure?

In such a world, directed work is impossible. The engine might spend energy furiously, but the result would be indistinguishable from staying still. There would be no reliable connection between doing *this* and getting *that*. The *Kriterion*'s discrimination would be useless, because the environment would not honor the distinction. The boundary would degrade at a rate set entirely by chance, and no amount of striving could alter that rate. Active engines, in a universe without causal regularities, could not arise, and if they did, they would dissolve instantly.

This is not an empirical observation about the dangers of chaos. It is a logical point about the concept of agency. Agency is the capacity to redirect causal flow. But "redirect" implies a direction—a difference that the agent's intervention makes to what would otherwise have occurred. If the universe provides no stable relation between interventions and outcomes, then the concept of making a difference collapses. Everything is a roll of the dice, and no player can ever learn to load them.

19.2 Minimal Determinism Defined

The required condition is *minimal determinism*. The term is chosen carefully. It does not mean hard determinism—a clockwork universe where every event is rigidly necessitated by prior states—though hard determinism certainly satisfies the condition. It does not rule out genuine indeterminacy at the quantum level, or probabilistic laws, or chaotic systems whose sensitivity to initial conditions limits long-term prediction while leaving their underlying dynamics fully law-governed. What it requires is that the world possesses *some* stable causal regularities—that there are patterns in the way events unfold, such that the same action, performed in the same context, tends to produce the same outcome, or at least shifts the probability distribution of outcomes in a consistent way.

Crucially, both hard determinism and probabilistic laws are instances of minimal determinism. The framework is indifferent between them. It does not need the universe to have "free will" in the sense of uncaused causes; it needs only that causal laws are stable enough to be exploited. As we will see, hard determinism is not a threat to agency but its firmest ground.[1]

[1]For the full argument on the deterministic nature of reality, see the forthcoming companion volume, *The Living Refutation: The End of the Determinist–Indeterminist War.*

19.3 The Argument from Existence

The argument is simple:

1. Active engines exist. We know this because we ourselves are active engines; we observe bacteria, plants, and animals that persist by redirecting causal flow.
2. Active engines can exist only if the universe possesses minimal determinism—enough causal regularity that actions can have reliable consequences.
3. Therefore, the universe possesses minimal determinism.

This is not a proof that the universe is fully deterministic, or that all events are predictable. It leaves open the possibility of irreducibly chancy processes, as long as those processes exhibit stabilities at the scale at which agents operate. A quantum world with probabilistic wavefunction collapse is compatible with minimal determinism, provided the probabilities themselves are stable and can be exploited. The rabbit need not know whether the fox's attack is deterministic or merely highly probable; it only needs that fleeing reduces the probability of being eaten by a positive margin.

What is revealed here is a profound "causal decoupling." This is not a decoupling from the laws of physics—an impossibility—but a decoupling from passive causal flow. A stone on a hillside is a slave to the ambient forces acting upon it; its trajectory is the sum of external vectors. An agent, however, imports energy from its environment and then spends that energy to bias outcomes toward its own persistence. By interposing its own *Kriterion-Organon* loop between environmental input and physical output, the agent decouples its fate from the default path of degradation. It steers. It uses the regularity of the universe as a lever to pry its own existence away from the statistical certainty of dissolution.

The argument is also not a proof that minimal determinism holds *everywhere* in the universe, or at all times. It holds wherever agency is possible. If there are regions of pure chaos—utterly lawless domains—then agency cannot exist there. But agency *does* exist here, so the domain of minimal determinism extends at least as far as the reach of any agent's actions.

19.4 What Must Be Ordered?

For agency to be possible, what exactly must be predictable? Not everything. An agent does not need to predict the weather a century hence, or the exact trajectory of every grain of sand in a desert. It needs to predict the consequences of its own actions *within the timescale and spatial range relevant to boundary maintenance.*

The bacterium must be able to rely on the fact that swimming in the direction of increasing nutrient concentration is more likely to bring it to food than swimming in a random direction. It does not need to predict the exact path of every nutrient molecule. The rabbit must be able to rely on the fact that fleeing into a burrow makes it harder for the fox to follow than standing still. It does not need to predict the fox's exact lunge trajectory.

The regularities that matter are those that couple actions to boundary-relevant outcomes. The world must be such that the *Kriterion* can learn, over evolutionary or individual time, a mapping from sensory states to effective actions. That mapping is the "*Logos*" of the agent's local environment—the structure that makes navigation possible. Minimal determinism is the guarantee that there is a *Logos* to learn.

19.5 The Link to the Boundary Integral

The boundary integral provides the formal statement of the requirement. Persistence is:

$$A(t) = A_0 + \int (\Phi - N)\, d\tau$$

For Φ to be positive—for directed work to have any meaning—there must be a stable causal relationship between the agent's energy expenditure and the values of Φ that result. In a purely random world, the integral of Φ over time would have an expected value of zero, regardless of the agent's actions. The boundary would degrade according to N alone, and no compensation would be possible. Minimal determinism is the condition under which the expected value of $\int \Phi\, d\tau$ can be raised by acting one way rather than another.

Thus, the very possibility of a positive boundary margin depends on the universe having a causal grain. Agency is the act of aligning with that grain. Persistence is the proof that alignment is possible.

19.6 Implications: Science, Learning, and the Navigability of Reality

The mandate of minimal determinism has profound consequences beyond biology. It grounds the very possibility of science and learning.

If agency requires exploitable regularities, then any agent capable of generalizing—of building abstract models—can in principle discover those regularities. The success of the natural sciences is not a lucky accident; it is a reflection of the same structural condition that makes life possible. The universe is arranged such that patterns can be detected and exploited. The *Logos* that the bacterium navigates with its chemoreceptors is the same *Logos* that the physicist maps with equations. The scale and abstraction differ, but the underlying condition—a world that answers to consistent laws—is the same.

Moreover, the mandate implies a kind of cosmic optimism about the limits of knowledge. Not that everything is knowable—there are domains beyond the reach of any finite agent—but that the world is not fundamentally deceptive. It does not present stable patterns for eons and then arbitrarily reverse them. An agent that invests in learning the causal structure of its environment is making a rational investment, because that structure is real and persistent. The rabbit that learns which sounds mean foxes and which mean wind is not a fool chasing phantoms. It is an agent exploiting the most basic promise of existence: that the world makes enough sense for sense-making to be worth the effort.

19.7 Clarifications and Forward References

The claim that the existence of active engines proves minimal determinism invites several challenges: that quantum indeterminacy undermines all regularity, that chaotic systems make agency impossible, that Humean skepticism defeats induction, that the argument proves only local order, and that hard determinism eliminates genuine agency. A brief signpost here removes immediate obstacles; the full steel-manned treatment belongs to Part VI (see Chapter 29: The Fourth Wall: Framework-Level Objections).

Quantum indeterminacy. Minimal determinism requires stable regularities at the scale of agency, not micro-level determinism. Quantum probabilities themselves are lawlike and reliable, and agents exploit statistical regularities just as effectively as deterministic ones.

Chaotic systems. Chaos limits the timescale of prediction but does not eliminate the underlying law-governed dynamics. Agents operate on the relevant short timescales, and life often exploits chaos (evasive prey trajectories) as much as it relies on order.

Humean skepticism. The framework does not solve Hume's problem of induction; it bypasses it with a transcendental argument: persistent agents exist, therefore the world is inductively tractable enough for agency.

Local order. The mandate requires only that the local environment of any agent is minimally deterministic. Global order is an empirical discovery, not a logical necessity.

Hard determinism and agency. Determinism is not the enemy of agency but its foundation: an active engine uses deterministic laws to bias outcomes toward persistence. The rock is determined and helpless; the bacterium is determined and effective.

19.8 The Universe That Agency Demands

We have arrived at a remarkable result. The very existence of life entails something about the kind of universe we inhabit. It is not a universe of pure flux. It is a universe with a grain, a *Logos*, a structure that can be learned and leveraged. The bacterium's swim, the rabbit's leap, the human's inquiry—all are the same act of aligning with that structure. They are possible only because the universe permits alignment.

Minimal determinism is not a law we discover by looking outward. It is a condition we prove by simply existing. Every agent is a walking demonstration that the world makes enough sense to sustain agents. The boundary that persists is the proof of the order that permits persistence.

Chapter 20

Nested Architecture and Conflict Resolution

Life is not a single boundary. From the very first protocell that engulfed a smaller replicator to the trillion-celled human reading this page, living systems have been built by stacking boundaries within boundaries. A cell is a bounded system maintaining its membrane; a tissue is a bounded system of cells held together by intercellular signals; an organ is a bounded system of tissues; an organism is a bounded system of organs. And beyond the organism, further boundaries appear—families, tribes, nations, ecosystems—each constituting a higher-order entity with its own *Hormē*.

This nesting is not a contingent architectural choice. It is the diachronic consequence of evolution, where individual agents, by cooperating or subsuming one another, create new, larger agents whose persistence depends on the coordinated *Hormē* of their parts. But with nesting comes a structural inevitability: the goods of the parts and the good of the whole can diverge.

20.1 The Logic of Nested *Hormē*

Recall that *Hormē* is the constitutive imperative of a bounded system to perform directed work for its own boundary maintenance. Proposition 10 showed that "good" for a system is whatever fulfills that *Hormē*; "bad" is whatever frustrates it. In a nested architecture, each boundary level defines its own good, because each boundary has its own *Kriterion* loop referencing its own integrity.

A cell's *Kriterion* tracks ion gradients, membrane tension, and nutrient availability. Its good is to maintain those parameters. A liver's cells collectively form a tissue boundary, whose good is to perform metabolic filtration and store glycogen for the organism. The organism itself—the highest-order boundary in the individual—has a good that encompasses the coordination of all its tissues and organs to maintain the macro-boundary (the skin, the immune system, the behavioral integrity of the whole).

Now consider a conflict. A liver cell, pursuing its own good, might detect a local shortage of energy and begin breaking down stored fat for its own use. This is, from the cell's perspective, good—it maintains its own membrane potential. But if all liver cells did this simultaneously, the organism's blood sugar would crash, threatening the brain and the macro-boundary's persistence. The cell's good and the organism's good pull in opposite directions.

This is not a flaw in the system. It is a direct consequence of architecture. Each boundary is operationally closed relative to its own *Kriterion*, and it cannot automatically "see" the consequences of its actions on the larger whole. The larger whole must therefore evolve mechanisms to constrain, coordinate, and sometimes override the lower-order *Hormē*.

20.2 The Inevitability of Internal Conflict

As systems grow in complexity, the primary threat to persistence often shifts from external disruption to internal civil war. An organism that can perfectly defend itself against predators and pathogens may still be destroyed by cancer: a lower-order boundary (the tumor) whose *Hormē* has decoupled from the organism's, replicating without limit until it consumes the host. A human mind can navigate social landscapes brilliantly, yet be undone by an addiction that hijacks the brain's reward circuitry into a local loop that treats drug-seeking as the highest good, even as the macro-boundary of health and relationships collapses.

These conflicts are not pathological exceptions. They are the expected operating condition of any sufficiently nested system. The very act of building a higher-order boundary out of lower-order agents means that those agents retain their own *Kriteria* and *Organa*; they are not passive cogs but active engines in their own right. The collective *Hormē* must be continually negotiated.

20.3 The Evolution of Higher-Order Governors

The solution—inevitable in any lineage that achieves stable nesting—is the evolution of higher-order *Kriterion* loops that monitor and modulate lower-order loops. These governors do not replace the lower *Kriteria*; they sit above them, receiving signals about the state of the whole, and exerting inhibitory or excitatory influence to bias lower actions toward the global good.

In human physiology, the prefrontal cortex (PFC) is the paradigmatic higher-order governor.[1] It does not directly control breathing or digestion—those lower loops operate independently. But when a lower impulse (e.g., a burst of anger from the amygdala, a craving from the nucleus accumbens) threatens the social boundary or long-term health, the PFC can suppress the impulse, shifting the balance of action toward a strategy that maximizes the whole organism's $A(t)$ over time. The felt experience of this conscious oversight is *self-control*; the felt experience of its failure is *akrasia* (ἀκρασία)—knowing the good and doing the opposite.

The architecture scales across the tree of life. A multicellular organism's immune system is a higher-order governor that detects rogue cells (cancer, infected cells) and eliminates them for the good of the body. A social insect colony has governors—pheromonal signals, worker policing—that suppress individual reproduction in favor of the queen's. A human society has laws, norms, and enforcement mechanisms that constrain individual *Hormē* to prevent the tragedy of the commons.

20.4 The Fourfold *Psyche*

The human mind is a nested architecture of *Hormē*, organized into four distinct but interacting strata. Each stratum is a system of *Kriterion*-action loops with its own reference states, its own timescale, and its own mode of operation. Together they form the *Psyche*, the total psychological boundary of the individual.

[1] Earl K. Miller and Jonathan D. Cohen, "An Integrative Theory of Prefrontal Cortex Function," *Annual Review of Neuroscience* 24, no. 1 (2001): 167–202, `https://doi.org/10.1146/annurev.neuro.24.1.167`.

The *Orexis* (Instinctual Layer). The lowest and most ancient stratum comprises the immediate biochemical drives for survival and homeostasis: hunger, thirst, pain avoidance, thermoregulation, sexual desire. These loops are hardwired and operate largely below conscious awareness. Their reference states are physiological set-points; their *Kriteria* are brainstem and hypothalamic circuits; their *Organa* are autonomic and endocrine responses. The *Orexis* demands immediate satisfaction and is indifferent to long-term consequences or social norms. It is pure, present-tense *Hormē.*

The *Thymos* (Social Layer). Above the *Orexis* sits the strategic social stratum, managing in-group/out-group dynamics through the social emotions: shame, pride, anger, loyalty, envy, righteous indignation. These loops track the boundary of the social self—one's standing, reputation, and belonging within a group. Their *Kriteria* evaluate one's status relative to others; their *Organa* produce emotional displays, social signaling, and coalitional behavior. The *Thymos* can override the *Orexis* (a hunger strike for a cause) and can be overridden by higher strata (suppressing anger to maintain a strategic relationship).

The *Logistikon* (Autopilot Layer). This is the domain of habit, routine, and learned automaticity. Through repetition and practice, actions that initially required conscious effort become encoded in procedural memory and can be executed without cognitive oversight—walking, driving, typing, playing an instrument. The *Logistikon* is a vast library of trained comparison-action loops that run on autopilot, freeing the conscious mind for higher-level tasks. Its reference states are learned expectations: how a familiar room should look, how a tennis swing should feel, how a social script should unfold. When the environment matches the learned pattern, the *Logistikon* executes smoothly and efficiently. When a mismatch occurs, it triggers an alert that calls the *Nous* into action.

The *Nous* (Conscious Mind). The highest stratum is the seat of reflective awareness, abstract thought, and long-term planning. The *Nous* constructs explicit models of the world, the self, and the future; it simulates possible actions and their consequences before committing to them. It is the only stratum capable of deliberative choice—the conscious commitment to one course of action over another based on a holistic assessment of the *Psyche*'s overall good. Crucially, the *Nous* is the *trainer* of the *Logistikon*: through deliberate practice and conscious repetition, it encodes new habits into the autopilot, gradually delegating routine tasks so that the *Nous* can attend to more complex challenges. The *Nous* does not eliminate the lower drives; it governs them by recalibrating their reference states, inhibiting their discharges, and, over time, rewiring the *Logistikon* to serve the whole.

This fourfold architecture resolves the earlier confusion about the "rational mind." The *Nous* is not a disembodied reason battling the passions; it is the highest-order *Hormē* loop, distinguished by its ability to model itself, its nested parts, and its diachronic trajectory. The *Logistikon*, far from being the rational executive, is the obedient automaton—indispensable for efficiency, but incapable of wisdom without guidance from the *Nous*.

20.5 Conflict as the Mother of Complexity

It is tempting to view internal conflict as a design flaw. The framework suggests the opposite. Conflict is the driving force behind the evolution of consciousness itself. A system with no internal conflicts—a completely homogenous agent with a single *Kriterion* and a single goal—needs no *Nous*. It simply does what its loop dictates. But a system composed of many sub-agents, each with its own *Hormē*, *must* evolve a higher-order arbiter, or it tears itself apart.

The expansion of the prefrontal cortex in hominids was likely driven not by the demands of the external environment alone but by the demands of managing an increasingly complex internal and social world, where the interests of the *Orexis*, the *Thymos*, and the *Logistikon* constantly collide. The *Nous* is the organ of internal diplomacy. Its felt experience of resolving a conflict between a bodily urge and a social norm—of saying "no" to a craving because it would jeopardize a relationship or a long-term goal—is the subjective signature of a higher-order loop successfully governing its lower parts.

This turns the traditional picture of reason and emotion on its head. The lower strata are not irrational; they are perfectly rational relative to their own narrow reference states. The craving for sugar is a flawless *Kriterion* loop: detect low blood glucose, trigger appetite, consume sugar. Its rationality is only *local.* The *Nous* provides *global* rationality—the integration of all local goods into a coherent, diachronic whole. The task of self-governance is not to extirpate desire but to train the *Logistikon* and recalibrate the *Thymos* so that their automatic outputs align with the long-term flourishing of the whole *Psyche.*

20.6 Pathology as Governance Failure

Many diseases can be reframed as failures of nested governance, with distinct failure modes at each stratum.

Addiction is a hijacking of the brain's reward circuitry—a lower *Orexis* loop that has artificially recalibrated its reference state to treat a drug as the supreme good. The *Logistikon* learns the addictive routines and executes them automatically. The *Nous* may retain an accurate cognitive model of the harm, but its inhibitory projections to the lower loops are weakened; the habit system runs unchecked. Treatment requires re-strengthening the *Nous*'s oversight and retraining the *Logistikon* with new, healthier habits.

Phobias and anxiety disorders reflect a miscalibration of the *Thymos* or *Logistikon.* A fear originally adaptive in an ancestral environment becomes encoded as an automatic response (*Logistikon*) or a social-evaluative hypervigilance (*Thymos*). The *Nous* knows the fear is disproportionate, but the autopilot fires before conscious oversight can intervene. Therapy involves re-exposing the system to the stimulus under safe conditions, allowing the *Nous* to update the *Logistikon*'s pattern-recognition.

Autoimmune disease is a failure of the immune governor to discriminate self from non-self; a lower boundary attacks the very whole it is part of.

Social fragmentation—factionalism, polarization, civil war—is a failure of higher-order social governors to maintain the legitimacy of the collective boundary. The *Thymos* of subgroups detaches from the overarching identity, and local goods overrule global goods.

In every case, the framework provides a diagnostic: identify the nested boundaries involved, trace their *Kriterion* loops, and locate the point where a lower loop's good has decoupled from the higher loop's good.

Treatment—whether medical, psychological, or political—consists of restoring the coupling, by recalibration, top-down inhibition, or structural realignment.

20.7 Clarifications and Forward References

The nested architecture of *Hormē* invites several challenges: that a hierarchy of *Kriteria* leads to an infinite regress (who governs the *Nous*?), that the higher-order good threatens to tyrannize lower strata, and that the binary dissolution threshold is too sharp for systems that degrade gradually. A brief signpost here removes immediate obstacles; the full steel-manned treatment belongs to Part VI (see Chapter 29: The Fourth Wall: Framework-Level Objections).

The homunculus regress. The regress stops at the outermost boundary of the organism. The *Nous* reflects on itself through meta-cognition, but the ultimate arbiter is the environment: if the *Nous* consistently misjudges, the organism dissolves.

Tyranny of the whole. The *Nous* does not dominate lower strata but integrates them. A healthy *Psyche* is a harmony, not a dictatorship.

Gradual dissolution. Boundaries can decline over years before crossing A_{min}. The framework distinguishes the direction of change from the moment of final collapse.

20.8 The Political Animal

Aristotle called the human a political animal.[2] The framework gives this a *Kybernēsis* foundation: to be a complex agent is to be a community of agents, each with its own *Hormē*, held together by a shared boundary and a multi-layered governance architecture. The *Psyche* is the internal *polis*: *Orexis* the populace, *Thymos* the factions, *Logistikon* the trained bureaucracy, and *Nous* the reflective statesman. Like any *polis*, its health depends on the wisdom of its leader, the loyalty of its functionaries, and the equilibrium of its competing interests.

[2]Aristotle, *Politics*, trans. C. D. C. Reeve (Hackett Publishing, 1998), 1253a

In the next part, we will trace how this architecture, derived here from first principles, was gradually uncovered by empirical science—from Schrödinger's physics to Friston's mathematics. The historical markers confirm that what the deduction proves is necessary, the laboratory finds to be actual.

Part V

Historical Integration

Chapter 21

Physics and Chemistry: Schrödinger, Prigogine, Gánti

A deductive argument that cannot explain the way things actually are is not philosophy; it is poetry. If the derivation of *Hormē*, agency, and value is correct, we would expect it to converge with empirical science at every point where science observes life at the highest resolution. The architecture the deduction makes mandatory should be the architecture the laboratory finds.

And it is. For nearly a century, physicists, chemists, and theoretical biologists have been uncovering fragments of that architecture—not because they were following a deductive trail, but because their empirical tools forced them to see what the logic of the boundary demands. They observed in nature what the derivation proves is structurally necessary. But because they built their theories upward from observation, they consistently encountered the mechanics without recognizing the mandate. They captured pieces of the puzzle without seeing the full picture that gives those pieces their coherence.

This chapter examines three foundational contributions from the physical and chemical sciences: Erwin Schrödinger's concept of negative entropy, Ilya Prigogine's far-from-equilibrium thermodynamics, and Tibor Gánti's chemoton model. Each marks a step closer to the deductive logic, and each reveals, by its very limitations, why a purely empirical approach cannot reach the constitutive truths the derivation delivers.

21.1 Erwin Schrödinger and the Thermodynamics of Order

In 1944, the physicist Erwin Schrödinger published a slim volume titled *What Is Life?*.[1] It was an audacious question for a physicist to ask, and Schrödinger answered it with a phrase that would echo through biology for decades: living organisms feed on "negative entropy." While the inorganic world drifts toward disorder and equilibrium, living things maintain their intricate organization by drawing order from their environment—consuming structured matter and excreting waste.

Schrödinger's insight maps cleanly onto the first two propositions of the derivation. He recognized that a living entity is a bounded, organized system (Proposition 1) and that the natural tendency of the physical world is to degrade that organization (Proposition 2). The organism must constantly offset this degradation by importing order from outside. The boundary, in Schrödinger's picture, is the frontier across which negative entropy flows.

What Schrödinger did not see—because he remained within the conceptual framework of thermodynamics—was the *directed* character of that flow. Negative entropy, in his account, arrives through metabolism: the organism eats, digests, and assimilates. But metabolism itself is not a passive soaking-up of order; it is an active, *Kriterion*-driven process. The organism does not merely absorb negative entropy; it *works* to extract it, discriminating between what is useful and what is poisonous, expending energy to channel ordered matter into boundary-maintaining pathways.

For Schrödinger, the "how" of this discrimination remained a black box. He knew that organisms maintained themselves against entropic decay, but he lacked the cybernetic vocabulary to describe the feedback loop that makes this possible. The constitutive imperative—the *must* of *Hormē*—is absent from his account. The organism, in his picture, persists because it happens to obey the laws of thermodynamics in a clever way. The derivation adds the missing piece: it persists because it is the kind of entity that *must* strive, or dissolve. Thermodynamics is the special case; the logical mandate is the general law.

[1] Erwin Schrödinger, *What Is Life? The Physical Aspect of the Living Cell* (Cambridge University Press, 1944).

Schrödinger's work is thus a milestone, not a destination. He correctly identified the physical problem that life must solve—continuous entropic degradation—and the general strategy—importing order. But he left the operational mechanics of that strategy unexplored, and he never asked what the very existence of such a strategy implies about the nature of agency and value.

21.2 Ilya Prigogine and Dissipative Structures

Three decades after Schrödinger, the Belgian physical chemist Ilya Prigogine won the Nobel Prize for demonstrating that far-from-equilibrium systems can spontaneously self-organize into coherent structures.[2] Hurricanes, convection cells, chemical clocks, and the Belousov-Zhabotinsky reaction all exhibit the same principle: when energy flows through a system at a high enough rate, ordered patterns emerge and persist as long as the gradient is maintained. Prigogine called these "dissipative structures," and he suggested that life itself is the most complex expression of this ordering principle.

Prigogine's work provides a powerful empirical vindication of Proposition 3—the Primal Distinction between passive conduits and active engines. Dissipative structures are not passive; they are organized patterns that harness energy flow to maintain themselves, at least for a time. A hurricane draws on the temperature gradient between warm ocean and cold upper atmosphere and, by doing so, sustains its characteristic spiral. This is not mere inertness; it is a kind of dynamic persistence.

But the hurricane stops precisely where life begins. The hurricane does not *redirect* the causal flow toward its own persistence; it *is* the causal flow. The pattern is the visible shape of the gradient's equalization. When the gradient dissipates, the hurricane vanishes without a trace of resistance. There is no *Kriterion*, no reference state tied to the hurricane's own boundary integrity, no constitutive imperative to remain a hurricane. The system is organized entirely by external forces; its organization is a transient eddy in the stream of energy, not an active engine pushing against the stream.

[2]Grégoire Nicolis and Ilya Prigogine, *Self-Organization in Nonequilibrium Systems: From Dissipative Structures to Order Through Fluctuations* (Wiley, 1977).

Prigogine's framework cannot, by itself, distinguish between a flame and a bacterium. Both are dissipative structures. Both consume energy and maintain a recognizable form. The difference—the *Kriterion* loop that biases energy deployment toward self-maintenance rather than self-extinction—is invisible to thermodynamics. Prigogine gave us the physics of the far-from-equilibrium, but he could not give us the *Kybernēsis* of the comparison-action loop. His active engines are missing the engine's governor.

The derivation completes Prigogine's picture by identifying what more must be added: a boundary that is actively maintained (not just externally imposed), a *Kriterion* that tracks the boundary's integrity, and a constitutive imperative that drives the loop. Thermodynamics supplies the stage; the deduction supplies the agent.

21.3 Stuart Kauffman and Autocatalytic Closure

In the 1990s and 2000s, the theoretical biologist Stuart Kauffman advanced the concept of autocatalytic sets: networks of molecules that mutually catalyze each other's formation, achieving a kind of collective self-reproduction.[3] Kauffman argued that such sets could emerge spontaneously from sufficiently diverse chemical mixtures, and that the property of "collective autocatalysis" constitutes a threshold for life. Together with the idea of a containing boundary—a membrane or coacervate—Kauffman's autocatalytic sets approach the minimal organizational closure that the derivation identifies as necessary.

Kauffman's contribution is significant because it operationalizes the boundary precondition (Proposition 1) at the chemical level. An autocatalytic set enclosed in a membrane is a bounded system whose interior reactions collectively sustain the membrane, and whose membrane, in turn, confines the reactions. This is a physical instantiation of the codependence of boundary and internal dynamics. Kauffman further emphasized that such systems are "holistic" and that their properties are emergent, not reducible to the properties of individual molecules.

But Kauffman's account, like Prigogine's, lacks the *Kriterion*. An autocatalytic set can sustain itself only as long as its environment supplies the right precursor molecules at the right rates. It has no mechanism for

[3] Stuart A. Kauffman, *Investigations* (Oxford University Press, 2000).

discriminating between favorable and unfavorable conditions, no sensor-*Organon* loop that actively seeks out resources or avoids poisons. Its persistence is a matter of chemical fortune, not directed work. It is a passive conduit at the molecular scale, albeit one with a beautifully intricate internal dance.

Moreover, Kauffman's reliance on "emergence" as an explanatory principle exposes a philosophical limitation. To say that life "emerges" from chemical complexity is to name a mystery, not to solve one. The derivation replaces emergence with deductive necessity: it shows exactly which additional features must be present for a chemical system to become an active engine, and why those features are mandatory, not miraculous. Kauffman provides a chemical mechanism; the derivation provides the logical architecture into which that mechanism must fit.

21.4 Tibor Gánti and the Chemoton

The closest historical approximation to the deductive method—and the most complete pre-echo of the present framework—is the chemoton model of the Hungarian biologist Tibor Gánti.[4] Working in relative isolation in the 1970s, Gánti set out to deduce the minimal abstract model of life, independent of any particular chemical substrate. The result was the chemoton: a system composed of three stoichiometrically coupled subsystems.

First, a **metabolic cycle** that takes in nutrients and produces both the materials for the other subsystems and its own catalysts. Second, a **template replication cycle**—a polymer that carries information and is copied by the metabolic cycle's products. Third, a **membrane boundary** that encloses the whole, regulating what enters and leaves, and that grows in concert with the internal cycles. The three subsystems are mutually dependent: the membrane contains the metabolism; the metabolism builds the membrane; the template organizes the metabolism.

Gánti's chemoton is a stunningly exact instantiation of Propositions 1 (the membrane boundary), 2 (disruption is inevitable, requiring constant renewal), 4 (the metabolic cycle performs directed work), and 8 (the loop

[4]Tibor Gánti, *The Principles of Life* (Oxford University Press, 2003).

runs continuously). Gánti even recognized that the chemoton must maintain a far-from-equilibrium state, and that its organization is the definition of its life. He did not rely on thermodynamics as the foundation; he reasoned from functional necessity.

Because Gánti's approach was deductive rather than empirical, he avoided many of the traps that ensnared later theorists. He did not attempt to derive life from chemistry by statistical emergence; he asked what chemistry must do to constitute a living system, and then looked for chemical implementations. His chemoton remains a touchstone for origins-of-life research, and it is often cited as the most rigorous minimal definition of life ever proposed.

And yet, even Gánti stopped short of the full derivation. The chemoton model describes a physical architecture; it does not articulate the constitutive imperative that drives it. Gánti says the chemoton works; the derivation says the chemoton *must* work, and that this "must" is the foundation of agency and value. The chemoton has no explicit *Kriterion*; its regulatory functions are distributed and implicit. Gánti did not extract the cognitive minimum, nor did he draw the conclusion that life and minimal agency are isomorphic. His model captures the *mechanics* of *Hormē* without naming the imperative.

Moreover, Gánti remained within the sphere of minimal biological definition. He did not extend the logic to epistemology, ethics, or the collapse of the Is-Ought gap. His chemoton answers the question "What is life?" but does not ask "What does life entail for value?" The present derivation begins where Gánti left off, showing that the very architecture he deduced contains within it the grounding of objective good.

21.5 The Pattern in the Fragments

Schrödinger, Prigogine, Kauffman, and Gánti were not working in the same tradition, and they did not speak a common language. But read together, and read through the lens of the eleven-step derivation, they tell a coherent story. They each grasped a piece of the puzzle.

Schrödinger saw the boundary and its vulnerability to entropy. Prigogine saw that order can self-organize far from equilibrium, and that life belongs to this family of ordered flows. Kauffman saw that chemical closure and autocatalysis can create the minimal organizational boundary. Gánti saw that the pieces—metabolism, information, container—must be stoichiometrically coupled into a single, self-sustaining whole.

What they all lacked was the *Kybernēsis* insight: the *Kriterion.* Without the *Kriterion,* the system is blind. Without blind striving, there is no directed work. Without directed work, there is no agency. And without agency, there can be no objective value. The empirical sciences of physics and chemistry can describe the hardware of life in exquisite detail. They cannot see the software—the continuous comparison-action loop that turns matter into an agent.

The derivation supplies the missing logic. It explains *why* the fragments must fit together as they do, and it reveals that the fitment itself—the mutual dependence of boundary, metabolism, and information—is not just a clever arrangement of molecules. It is the logical consequence of being a determinate entity in a disruptive world. The pieces are not accidental; they are mandated. Physics and chemistry provided the observational data; the deduction provides the philosophical architecture that makes the data intelligible.

The next chapter moves from the physical sciences to the life sciences proper. Physiologists and cyberneticians like Claude Bernard, Walter Cannon, and W. Ross Ashby discovered the homeostatic loop in living organisms, confirming Propositions 5 through 8. But they, too, missed the deeper imperative, conflating life with mere regulation. We turn to their story next.

Chapter 22

Cybernetics and Homeostasis: Bernard, Cannon, Wiener, Ashby

The physicists and chemists of Chapter 21 described the container and its fuel. They showed that life organizes matter far from equilibrium, that it imports order to offset decay, and that its minimal architecture is a self-sustaining chemical system enclosed by a boundary. But they did not describe the *steering*—the ongoing, moment-by-moment adjustments by which a living system keeps itself within the narrow range of viability. That discovery fell to the physiologists and, later, the cyberneticians. Their work confirmed Propositions 5 through 8 of the derivation in exquisite empirical detail. But it also introduced a philosophical blind spot that the framework alone can correct: the set-point error.

22.1 Claude Bernard and the Inner World

In the 1860s, the French physiologist Claude Bernard proposed a principle that would become the foundation of regulatory biology.[1] The living body, he argued, maintains a stable internal environment—the *milieu intérieur*—distinct from the external world. All the functions of the organism, however varied, serve a single end: "to preserve constant the conditions of life in the internal environment." Without this constancy, the delicate processes of cells and tissues would fail. Life is possible only because the interior is buffered against the chaos outside.

Bernard's insight is a direct empirical observation of the boundary precondition (Proposition 1) and the inevitability of disruption (Proposition 2). The boundary is not merely a spatial separation; it is a functional one. It distinguishes an internal chemical composition, temperature, and osmotic pressure from the external surround. And it is continuously under

[1] Claude Bernard, *Introduction à l'étude de La Médecine Expérimentale* (Baillière, 1865).

threat: even breathing changes the blood's acidity; even a slight change in external temperature seeps inward. The organism, by Bernard's account, must actively counteract these perturbations, or the *milieu intérieur* would drift into uninhabitability.

What Bernard did not specify was *how* the organism achieves this constancy. He described the result—stability—but not the mechanism. The concept of a *Kriterion* loop, of a sensor that measures a variable and triggers a corrective action, was still decades in the future. Bernard's organism is a wise steward of its inner world, but the steward's operating manual remains locked.

22.2 Walter Cannon and the Wisdom of the Body

Half a century later, the American physiologist Walter Cannon gave Bernard's principle a name: *homeostasis*.[2] Cannon demonstrated that the body's regulatory systems are not passive buffers but active, feedback-driven processes. When blood sugar falls, the pancreas secretes glucagon; when it rises, insulin. When body temperature drops, shivering and vasoconstriction are engaged; when it climbs, sweating and vasodilation. These responses are proportional to the deviation, and they counteract it, restoring the variable to its set-point.

Cannon's work is a physiological exemplar of Propositions 5 and 6. The body contains explicit *Kriteria*: cells in the hypothalamus that measure blood temperature, cells in the pancreas that measure glucose. Each *Kriterion* has a reference state—the set-point—and a mechanism for translating deviations into corrective action. The model (the set-point and the detection apparatus) and the action (the *Organon* response) are co-constitutive; one cannot meaningfully speak of the measurement without the response it triggers, nor of the response without the measurement that guides it.

Cannon's phrase "the wisdom of the body" captured the apparent intelligence of these processes. But the phrase is, in his hands, a metaphor. Cannon remained a physiologist; he described the mechanisms without asking why they are there, why they *must* be there. He took the set-points as given—natural constants of the organism—and did not inquire into their origin or their normative status. The wisdom of the body is, for Cannon, a mechanical marvel, not a constitutive imperative.

[2]Walter B. Cannon, *The Wisdom of the Body* (W. W. Norton & Company, 1932).

22.3 Norbert Wiener and the Science of Control

The next leap came not from biology but from engineering and mathematics. During and after the Second World War, Norbert Wiener developed the theory of *cybernetics*: the study of communication and control in animals and machines.[3] Wiener saw that the same principles of feedback that govern a thermostat or a target-tracking anti-aircraft gun also govern the nervous system of a living creature. Negative feedback—feeding the deviation from a goal back into the system as a corrective signal—is the universal logic of purposive behavior.

Wiener's cybernetics gave a rigorous mathematical language to the *Krite-rion*-action loop. A system receives sensory input, compares it to a reference value (the goal, the set-point), computes the error, and acts to reduce the error. This is Proposition 8: persistence is a continuous *Kybernēsis* process. And Wiener explicitly recognized that this loop could be instantiated in flesh, metal, or circuitry; the logic was substrate-neutral.

But Wiener, like Cannon, stopped at the mechanism. For him, a guided missile and a hungry cat were structurally equivalent. Both exhibited purposeful behavior; both corrected their course in response to feedback. The difference between them was merely one of complexity and material, not of kind. This flattening of the distinction between life and machine was a deliberate philosophical stance: cybernetics was meant to dissolve the boundary between the living and the artificial, to see both as information-processing systems.

The derivation reveals the flaw in this flattening. A guided missile's goal—its set-point—is externally imposed by its designers. The missile does not care whether it hits the target; it does not maintain its own boundary; it does not repair its own wiring. Its comparison-action loop is not constitutive of its own existence; it is a tool, an extension of the *Hormē* of the human agency that built and launched it. The cat's goal—to catch the mouse, to maintain its body temperature, to survive—is intrinsic. Its set-points are not arbitrary engineering choices; they are the conditions of its own continued existence. The cat *must* strive, or it dissolves. The missile, if it fails, simply becomes scrap.

[3] Norbert Wiener, *Cybernetics: Or Control and Communication in the Animal and the Machine* (MIT Press, 1948).

22.4 W. Ross Ashby and the Homeostat

If Wiener was the visionary of cybernetics, W. Ross Ashby was its most rigorous experimentalist. In the 1940s and 1950s, Ashby built a device called the *homeostat*: four interconnected electrical units that, when perturbed, automatically readjusted their internal states until they returned to equilibrium.[4] The homeostat had no set-point programmed into it; the set-point was defined only by the system's own dynamics—the stable configuration to which it spontaneously returned.

Ashby argued that the homeostat was a model of the brain and of life itself. It demonstrated that "purposeful" behavior—the active restoration of a stable state after disturbance—could arise from a purely physical system, with no need for consciousness, intention, or a designer's blueprint. The homeostat's behavior was *teleonomic* in the language of Chapter 15: Teleonomy Without Teleology: the apparent purpose was a consequence of the system's physical organization, not of a mental goal reaching backward in time.

Ashby's homeostat is a powerful vindication of Proposition 7—the constitutive imperative. The system *must* return to its stable state, not because it is commanded to, but because that is the only state in which its dynamics can persist. There is no external "should"; the imperative is built into the wiring. And Ashby, unlike Wiener, recognized that this self-referential closure was the heart of the matter: a truly autonomous system must define its own goals, not have them imposed.

Yet even Ashby did not draw the full philosophical conclusion. His homeostat lacks a boundary in the sense of Proposition 1. It is a circuit on a bench, plugged into the wall, maintained by a human experimenter. Its persistence depends on external power and repair. It does not maintain its own material integrity; if a capacitor blows, it does not solder a new one. The homeostat is a model of *internal* regulation, but not of *self-constitution.* It shows how a system can stabilize its own states, but not how a system can actively maintain the very boundary that separates it from the world.

[4] W. Ross Ashby, *Design for a Brain* (Chapman & Hall, 1952).

22.5 The Set-Point Error

The scientific lineage from Bernard to Ashby established, beyond reasonable doubt, that living systems are *Kybernētēs.* They sense, compare, and act to maintain internal states against external perturbation. The derivation, in Propositions 5 through 8, *deduced* that this architecture is mandatory; the physiologists and cyberneticians *observed* it in operation. The convergence of deduction and empirical discovery is a powerful corroboration.

But the cybernetic tradition carried a philosophical error that has persisted into contemporary cognitive science. By treating all negative-feedback systems as structurally equivalent, it conflated two fundamentally different kinds of set-point: the extrinsic and the intrinsic.

An extrinsic set-point is one imposed from outside the system by a designer or an external agent. The thermostat's target temperature is set by the human occupant. The guided missile's target coordinates are uploaded by the weapons officer. The chess AI's winning condition is programmed by the software engineer. These systems are tools; their purpose is derivative.

An intrinsic set-point is one that arises from the system's own boundary condition. The bacterium's chemotactic baseline, the human's body temperature, the cell's membrane potential—these are not chosen by anyone. They are the physical parameters that must be maintained for the system to continue existing as itself. They are the *constitutive* conditions of the system's identity. An intrinsic set-point is not a preference; it is a prerequisite.

The cybernetic tradition missed this distinction. It saw the loop but not the source of the reference signal. Consequently, it could not explain why living systems are *agents* while guided missiles are not. Both steer; both correct error; both appear purposeful. But only one steers for its own sake. The derivation supplies the missing criterion: intrinsic *Hormē*, the constitutive imperative that links the *Kriterion*'s reference state to the system's own continued existence.

This is not a mere semantic quibble. The set-point error is the reason cybernetics could never bridge from the physical description of feedback to the objective grounding of value. If a system's goals are always externally imposed, then all goals are arbitrary, and value is ultimately a matter of the designer's preference. If a system's goals can be intrinsic—if they are the conditions of that very system's existence—then value is anchored in the physics of identity. The derivation corrects the set-point error by identifying *Hormē* as the universal, constitutive reference: the boundary's integrity is the set-point that no living system can negotiate away.

22.6 From Mechanism to Mandate

The story of cybernetics and homeostasis is, in retrospect, a story of proximity to a truth that remained just out of reach. The physiologists and cyberneticians built the mechanical picture with extraordinary precision. They described the *Kriterion*, the *Organon*, the loop, the continuous regulation, the apparent purpose. They even, in Ashby's homeostat, glimpsed the self-defining nature of the set-point.

What they lacked was the deduction from first principles. Because they worked upward from observation, they could not see the *necessity* behind the machinery. They could not answer *why* a living system must be a *Kybernētēs*, *why* its loop must be continuous, *why* its set-point must be intrinsic. And without that deduction, they could not draw the normative conclusions—that life and agency are isomorphic, that value is objective, that truth and goodness are the same loop tracking the causal order.

The derivation completes their project. It does not discard the cybernetics; it grounds it. It shows that the loop is not an evolutionary trick that happened to work, but the logical consequence of being a bounded entity in a disruptive universe. The physiologists gave us the wisdom of the body; the derivation reveals that this wisdom is not metaphorical. It is the only possible strategy for a system that refuses to dissolve.

The next chapter turns to a parallel tradition in biology—the theory of autopoiesis and the concept of *Umwelt*—which came even closer to the isomorphism of life and agency, yet veered into a radical constructivism that the derivation must correct.

Chapter 23

Autopoiesis, Umwelt, and the Constructivist Trap

The cyberneticians described the loop. They showed that living systems maintain internal stability through continuous comparison and action, and they built machines that did the same. But they left open a question they could not answer: where does the loop's reference state come from? Cannon's set-points were taken as given. Wiener's goals were programmed in from outside. Ashby's homeostat found its own equilibrium, but only within a boundary maintained by the experimenter. None of them explained how a system could *generate* its own boundary and, in so doing, define its own norms.

That explanation came from biology, and specifically from a tradition that understood organisms not as machines but as *worlds*. The ethologist Jakob von Uexküll introduced the concept of the *Umwelt*—the species-specific perceptual universe in which an organism lives and to which it responds.[1] The biologists Humberto Maturana and Francisco Varela built on this insight to propose a radical new definition of life itself: autopoiesis, the self-production of a boundary.[2] Together, these thinkers came closer than any before them to the life-agency isomorphism at the heart of the derivation. They recognized that living is a kind of knowing, and that the boundary is the locus of both. But they drew a fateful conclusion: because all knowledge is bounded by the organism's own organization, objective knowledge of reality is impossible. This was the constructivist trap, and the derivation springs it.

---

[1]Jakob von Uexküll, *Streifzüge Durch Die Umwelten von Tieren Und Menschen: Ein Bilderbuch Unsichtbarer Welten* (Julius Springer, 1934).

[2]Humberto R. Maturana and Francisco J. Varela, *Autopoiesis and Cognition: The Realization of the Living* (D. Reidel, 1980).

23.1 Jakob von Uexküll and the Subjective Universe

In the early twentieth century, while behaviorists were reducing animals to stimulus-response machines, the German-Estonian biologist Jakob von Uexküll was doing something radically different. He was asking what the world looks like *from the perspective of the animal itself.* His answer was the *Umwelt*: not the physical environment as a human observer would describe it, but the environment as it exists *for* the organism, defined by the features that matter to its survival and action.

A tick, famously, climbs a grass stalk and waits. It has no eyes, no ears; its world is a dark silence. But it has a sensor tuned to butyric acid, a volatile compound exuded by mammals. When the acid hits the sensor, the tick drops. If it lands on a warm body, it finds a patch of skin and feeds. The tick's *Umwelt* is composed of exactly three stimuli: the smell of sweat, the heat of a body, the texture of skin. Everything else—the sun, the grass, the passage of seasons—does not exist for the tick, because the tick has no *Kriterion* to register it.

Uexküll's insight maps directly onto Proposition 5: the cognitive minimum. An organism's *Kriterion* is tuned to a specific set of environmental variables—those that are relevant to boundary maintenance. The tick does not need a full model of the forest; it needs a model of mammal-proximity. Its sensorium *is* that model, and the world it registers is the only world that matters. The *Umwelt* is the perceptual face of the *Kriterion*: the organism senses what it needs to sense, and the rest is silence.

Uexküll also understood that the *Umwelt* is co-constituted with action. The tick's world is not just a set of sensations but a set of *opportunities for action*—what later theorists would call affordances. The butyric acid signal does not merely inform the tick that a mammal is passing; it *releases* the dropping reflex. Sensing and acting are bound together in a single functional cycle, exactly as Proposition 6 describes: the model and the action are co-constitutive.

Uexküll's work is an almost perfect empirical illustration of the derivation's first nine propositions. The tick has a boundary, it is subject to disruption (it must feed or die), it discriminates, it acts, its loop is continuous during its active phase, and it exhibits a clear intrinsic *Hormē.* It is a minimal agent. And its agency is defined by the world it enacts.

But Uexküll, having discovered the *Umwelt*, drew a philosophical conclusion that the derivation will ultimately reject. If the world as experienced is always relative to the organism's sensorium, then there are as many worlds as there are species. Each *Umwelt* is a closed bubble of perception and action, and the bubbles do not touch a common reality. The tick's world and the human's world are mutually alien. There is no objective environment; there are only the countless subjective environments of countless organisms. The physical world, in this view, is an abstraction—perhaps useful, perhaps not—but it is not what any organism actually inhabits.

This is the seed of constructivism: the view that minds do not discover reality but construct it from their own organizational closure. Uexküll's observations were brilliant, but his epistemology was premature. The derivation shows that the *Umwelt* is not a bubble that isolates the organism from reality; it is a tunnel that connects the organism to the *specific causal features of reality that impinge on its boundary.* The tick's butyric acid receptor is not a subjective projection; it is a physical coupling to a real chemical. If that coupling is inaccurate, the tick dies. The *Umwelt* is selective, but what it selects is real.

23.2 Maturana, Varela, and the Logic of Self-Production

Half a century after Uexküll, the Chilean biologist Humberto Maturana and his student Francisco Varela took the next decisive step. In the 1970s and 1980s, they proposed that life should be defined not by reproduction, metabolism, or evolution, but by *autopoiesis*: the continuous self-production of the system's own boundary and internal components.[3]

An autopoietic system is a network of processes that produces the very components that constitute the network, and that also produces the boundary that contains the network. A living cell is the paradigm: its membrane, enzymes, and genetic machinery are all produced by the cell's own metabolic activity, and the membrane, in turn, provides the compartment within which that activity occurs. The system is operationally closed: it is defined by the circularity of its own organization, not by inputs and outputs. Inputs enter, outputs leave, but the organization—the

[3]Maturana and Varela, *Autopoiesis and Cognition.*

pattern of relations that constitutes the system—is maintained only by the system's own work.

Maturana and Varela recognized that autopoiesis is precisely what the derivation has called intrinsic *Hormē*. The system's *Kriterion* reference is its own boundary integrity; its set-points are the conditions of its continued self-production. An autopoietic system cannot be externally programmed to be what it is; its identity is its own ongoing achievement. This is the constitutive imperative (Proposition 7) stated in the language of systems biology.

Moreover, Maturana and Varela explicitly argued that cognition is not a separate function added on top of life. It is the very same activity. An autopoietic system, by maintaining its boundary against perturbation, is engaging in an ongoing process of discrimination and response that *is* cognition. "Living is knowing," they wrote. Bacteria know their chemical environment; immune systems know foreign proteins; nervous systems know patterns of sensation. There is no threshold at which cognition suddenly appears; it is present, in its minimal form, wherever autopoiesis operates.

This is the most explicit statement of the life-agency isomorphism (Proposition 9) in the scientific literature. Maturana and Varela saw, as clearly as any biologist has ever seen, that the boundary-maintaining loop *is* the cognitive loop, and that both are definitive of life. They did not need the eleven-step deduction to reach this conclusion; they arrived at it empirically, by studying what organisms actually do.

And yet.

23.3 The Constructivist Trap

From the operational closure of autopoietic systems, Maturana and Varela drew a conclusion that has become the orthodoxy of radical constructivism. If a system can only respond to perturbations in terms of its own internal organization—if its cognition is always a matter of compensating for disturbances, not of accessing an independent reality—then the system can *never know the world as it is.* It can only know the world as it appears through the lens of its own structure. The *Umwelt* is not a window onto reality; it is a prison.

Varela, in later work, described cognition as "enaction": the organism brings forth a world through its sensorimotor coupling, and that world is not a representation of a pre-given reality.[4] Though Varela explicitly sought a via media—a middle way to avoid the pure idealism of radical constructivism—the relentless logic of operational closure ensures that this enacted world ultimately collapses into an internal construction. There is no "view from nowhere," no objective standpoint from which to compare the organism's model to the territory. Knowledge is always knowledge-for-someone, and the someone is always bounded by its own organization.

This conclusion seems to follow naturally from the premises. If all cognition is autopoietic—if the only way a system can "know" is by compensating for perturbations relative to its own reference states—then it appears that we are trapped in a structural solipsism. We can never step outside our own nervous systems to check whether our models correspond to an independent reality. The very concept of objective truth is discarded in favor of a closed, relational matrix.

The constructivist trap is seductive because it builds on undeniable truths. It is true that our access to the world is mediated by our sensors. It is true that our brains construct models, and that those models are constrained by our neural architecture. It is true that we have no magical faculty of direct, unmediated intuition of reality. But it does not follow that we cannot know objective reality. It follows only that we cannot know it exhaustively, and that our knowledge is always provisional. The collapse from "knowledge is bounded" to "knowledge cannot mirror objective reality" is a non sequitur, and it reveals where the enactive tradition lost its nerve and retreated into the very idealism it sought to escape.

23.4 Breaking Out: The Objective Anchor

The derivation provides the missing anchor. Recall Proposition 11: the truth-goodness isomorphism. For an agent, epistemic error and practical failure are the same thing. A model that does not track the causal structure of reality will lead to actions that degrade the boundary. If the degradation is severe enough, the system dissolves.

---

[4] Francisco J. Varela et al., *The Embodied Mind: Cognitive Science and Human Experience* (MIT Press, 1991), p. 148

This is not a philosophical abstraction. It is a physical fact that holds for every autopoietic system. The bacterium's chemoreceptor is not a "construction" of the chemical environment; it is a physical molecule that changes shape when it binds a specific ligand. That change is caused by the real chemical. If the receptor were tuned to something that did not exist, it would provide no survival advantage; the bacterium would fail to find glucose and would starve. If the receptor were tuned to something that is actually poison, it would kill the bacterium. The only reason the receptor exists and persists across generations is that it tracks a real feature of the world—a feature that, on average, conduces to the bacterium's survival when detected.

Thus, the operational closure of an autopoietic system does not cut it off from reality. It *depends* on reality. The system's continued existence is an ongoing proof that its internal organization is, at some level of description, aligned with the causal structure of the outside. The alignment is not perfect; it need only be good enough to keep the boundary margin positive. But it is real. Reality is what the system is tracking when it survives.

The constructivists mistook the *selectivity* of cognition for its *fabrication.* The tick's *Umwelt* is selective—it ignores the sun, the grass, the earth—but the butyric acid it detects is real. The cell's membrane is selective; it lets some ions through and not others. But the gradients it maintains are real. Operational closure means that the system determines *how* it responds to perturbations, not that perturbations are imaginary. The system is not a solipsist; it is a specialist. It tracks what it needs to track, and the proof that it tracks accurately is that it persists.

The derivation thus completes the project that Maturana and Varela began. They correctly identified the architecture: life is autopoietic, cognition is built into that autopoiesis, and agency emerges from the loop. But they hesitated at the epistemic conclusion. They believed that because all knowledge is mediated, no knowledge is objective. The derivation shows that this is a mistake. Mediation does not imply fabrication. The *Kriterion* is a mediator, but it is a mediator that *must* be causally coupled to the world. When it is not, *Dikē* enforces the result. Reality does not care about your model of it; it simply imposes its order, and an inaccurate map dissolves.

23.5 Varela's Late Turn: The Emergence of Sense-Making

To be entirely fair to the enactive tradition, Francisco Varela himself eventually sensed the suffocating walls of the constructivist trap. In his later work, particularly in his collaborations with Andreas Weber, Varela realized that a purely mechanical, structurally determined description of autopoiesis was incomplete.[5] It could describe a system's *operation*, but it could not describe its *life*.

To bridge this gap, Varela introduced the concept of *"sense-making."* He argued that an organism is not merely a closed circuit of chemical reactions; it is a locus of perspective. By the very virtue of its autopoietic maintenance, the organism casts a horizon of significance over its environment. A molecule of glucose is not just a neutral chemical perturbation; to a bacterium, it is *good*. A toxin is not just a structural disruption; it is *bad*. Varela called this *"intrinsic teleology"*—the biological generation of a primitive, naturalized value system where things in the world are evaluated relative to the system's own continuation.

With this turn, Varela was walking directly toward the architecture of the *Kriterion*. He caught a glimpse of the truth that life and value are isomorphic. He saw that an organism must evaluate its environment as "better" or "worse" to guide its behavior.

And yet, Varela lacked the physics to anchor his intuition.

23.6 The Unfinished Bridge

Because Varela remained tethered to the phenomenological tradition, he framed sense-making as an *emergent, lived experience* rather than a strict logical constraint. He could assert that sugar was "good" for the bacterium, but because he still clung to the absolute operational closure of the system, he could not mathematically explain how that internal feeling of "goodness" maintained a true, objective correspondence with the external universe. He had identified the necessity of a navigational loop, but he had no way to compute the drift.

This is where the boundary integral completes Varela's unfinished journey. The *Kriterion* is not a poetic horizon of significance that an organism

[5] Andreas Weber and Francisco J. Varela, "Life After Kant: An Introduction to a Naturalistic Teleology," *Phenomenology and the Cognitive Sciences* 1 (2002): 97–125.

"brings forth" out of its own subjectivity. It is a strict physical requirement dictated by the hard math of survival:

$$A(t) = A_0 + \int_0^t (\Phi - N)d\tau$$

The organism's evaluation of the world as "good" or "bad" is not an idealist construction; it is the real-time tracking of whether an environmental feature decreases entropic dissipation (N) or scales up directed work (Φ).

By anchoring Varela's "sense-making" into the unyielding physical requirement of boundary preservation, we rescue his naturalistic teleology from the constructivist trap. The organism behaves as a center of significance because it faces an objective, quantifiable deficit. It does not just interpret a world; it calculates its own persistence within it.

23.7 Autopoiesis Meets the Boundary Integral

The boundary integral is the mathematical architecture that completes Varela's unfinished bridge. It formalizes his 'horizon of significance' into an unyielding mathematical constraint:

$$A(t) = A_0 + \int_0^t (\Phi - N)d\tau$$

For the autopoietic system, Φ is the metabolic work that maintains the membrane, and N is the entropic and environmental degradation. But Φ is not a blind force; it is directed work, guided by the *Kriterion*. The *Kriterion*'s accuracy determines how effectively Φ offsets N. If the *Kriterion* systematically misrepresents the environment—if it triggers feeding when poison is present—then $A(t)$ will drop, and the system will dissolve.

The fact that autopoietic systems persist for billions of years is therefore proof of an objective causal coupling between their *Kriteria* and the world. They persist not because they "construct" a reality that happens to work, but because their construction is selectively shaped by a reality that did not wait for their opinion. The environment kills the inaccurate. The survivors are the accurate. The constructivist argument that we can never *know* we are accurate is true only in the sense that we can never stand outside our models to verify them with certainty. But we can and do stand inside our models, and the world verifies them moment by moment, by the simple expedient of letting us continue to exist.

23.8 The Legacy and the Correction

The autopoiesis tradition, and its descendants in enactive cognitive science, made a permanent contribution. They showed that life and cognition are not separate; that mind is not a late-emerging ghost in a biological machine but the continuous activity of the machine itself; that agency is rooted in the very physics of self-maintenance. These are permanent insights, and the derivation affirms them as deductive truths.

But the tradition also spread a confusion that has done real harm. By insisting that all knowledge is constructed and that objective reality is inaccessible, it severed the link between truth and survival that is the bedrock of any coherent epistemology. It gave license to the idea that different *Umwelten* are incompossible—that there is no way to adjudicate between the model of the shaman and the model of the scientist, because all models are equally "enacted." The derivation cuts through this relativism. Models are not equal. Some lead to dissolution; some lead to persistence. The difference is objective, and it is measured in the continued existence of the agent.

This does not mean that all truth is reduced to brute survival. The truth-goodness isomorphism is a foundational conclusion, not a complete epistemology. Complex cognitive systems build models that far exceed the immediate demands of boundary maintenance—models of the cosmos, of abstract mathematics, of the distant future. But those models inherit their epistemic grounding from the same *Kriterion* architecture that keeps the cell alive. The pursuit of truth at the highest levels is an extension of the bacterium's search for glucose, and it is answerable to the same ultimate arbiter: *Dikē*.

23.9 The Next Fragment: Teleodynamics and Information

With the correction of the constructivist trap in place, the historical trajectory of the life sciences can be seen in a new light. The physiologists found the loop; the cyberneticians formalized it; the autopoietic theorists located it in the self-producing boundary and recognized it as cognition. At each stage, the empirical description converged more tightly with the deductive architecture. And at each stage, a philosophical error—the set-point error, the constructivist trap—prevented the final unification.

The next step in the historical progression would be taken by theorists who attempted to ground agency in thermodynamics, information theory, and the mathematics of predictive processing. Terrence Deacon, Giulio Tononi, and Karl Friston each developed frameworks that capture aspects of the *Kriterion* architecture with mathematical precision. But each also inherited the limitations of an empirical approach: they described the engine in ever finer detail without naming the constitutive imperative that turns the key. Their work, and its completion by the derivation, is the subject of the next chapter.

Chapter 24

Teleodynamics and Information: Deacon, Tononi, Friston

The autopoietic tradition identified the loop and the boundary. It recognized that life is the continuous self-production of the very container that houses it, and that this self-production is inherently cognitive. But autopoiesis remained, in certain respects, a qualitative description. It gave a picture of the dynamics but not a rigorous mathematical formalism that could allow prediction or measurement. The next generation of theorists—Terrence Deacon, Giulio Tononi, and Karl Friston—filled that gap. They developed frameworks that quantify, with increasing precision, the causal architecture of agents, the information flows that maintain them, and the predictive loops that keep them aligned with their environments. Between them, they capture the key elements of the derivation: the directedness of work, the continuity of the loop, and the self-referential nature of the *Kriterion*. But each framework remains fundamentally descriptive, unable to ground the normative force of the imperative it describes. They built the engine; they did not explain why it must run.

24.1 Terrence Deacon and the Dynamics of Absence

The American anthropologist and neuroscientist Terrence Deacon, in his 2011 book *Incomplete Nature*, asked a question that had haunted science since the seventeenth century: how can purpose exist in a universe of merely physical causes? Deacon's answer is *teleodynamics*—a higher-order dynamics that arises when two or more self-organizing processes become coupled such that they mutually constrain each other toward ends that are not locally present.[1]

[1]Terrence W. Deacon, *Incomplete Nature: How Mind Emerged from Matter* (W. W. Norton & Company, 2011).

Deacon begins with the familiar concept of *morphodynamics*: the tendency of matter, far from equilibrium, to spontaneously form coherent structures, such as snowflakes or convection cells. He then adds a second layer: *teleodynamics*, which occurs when morphodynamic processes become organized around a reference state that is *absent*—a state the system is not currently in but toward which its dynamics are directed. He calls this "absential causation": the causal power of something that is not there.

An organism maintaining homeostasis is a paradigm: it acts not because of the present state directly, but because the present state deviates from a norm, and that norm is what drives the action. The norm is physically absent—it exists only as a structural predisposition—but it causally shapes the system's behavior. Deacon argues that this absential causation is the physical basis of purpose. It is what distinguishes an organism from a whirlpool: both are processes, but the organism's processes are constrained by a reference condition that must be continuously re-established.

Deacon's teleodynamics maps remarkably well onto the derivation's logic. The reference state is the *Kriterion*'s set-point; the constraint of morphodynamics toward that state is directed work; the entire process is teleonomic, not teleological, in exactly the sense of Chapter 15. Deacon explicitly rejects final causes and grounds purpose in the architecture of self-organization. He even anticipates the constitutive imperative: a teleodynamic system must continue to organize itself toward its absent reference, or it ceases to be.

But Deacon's framework suffers from a foundational ambiguity. He presents teleodynamics as an *emergent* property—something that arises out of the proper arrangement of morphodynamic processes, and that does not exist at lower levels. Purpose, in his view, is a higher-level phenomenon that "emerges" when certain forms of complexity are reached. This reliance on emergence, however, leaves a critical question unanswered: *why* does emergence happen? If teleodynamics is just something that happens to appear when certain physical conditions are met, then purpose remains a contingent accident of cosmic history, not a structural necessity. The derivation fills this gap: the *Kriterion* loop and its constitutive imperative are not emergent mysteries; they are deductively necessary consequences of being a bounded entity in a disruptive world. Teleodynamics is not an emergent bonus; it is the only way a bounded system *can* persist. The "absent" reference is not a ghostly force; it is the physical condition of the boundary's integrity, encoded in the system's architecture by past persistence. Deacon correctly described the dynamics of *Hormē*; the derivation shows why those dynamics are mandatory, not optional.

24.2 Giulio Tononi and the Measure of Causal Power

If Deacon asked how purpose can exist physically, the neuroscientist Giulio Tononi asked a complementary question: how can a physical system be a *subject of experience*—a unified self that exercises causal power over its own states? His answer is Integrated Information Theory (IIT), which proposes that consciousness is identical to the amount of integrated information a system generates, quantified by a measure called Φ (Phi, not to be confused with the directed work of the boundary integral). More fundamentally for our purposes, IIT attempts to formalize what it means for a system to possess *intrinsic causal power*: the ability to make a difference to itself, above and beyond the sum of its parts.[2]

IIT begins with a deceptively simple move. Instead of describing a system from the outside, as an observer would, it asks: what does the system *specify*, from its own intrinsic perspective, about its possible past and future states? A system has intrinsic causal power if its current state, given its internal mechanisms, uniquely constrains both what could have caused it and what effects it can have. The more the system's mechanisms constrain these possibilities in an integrated and irreducible way, the higher its Φ, and the more consciousness it possesses.

Tononi's framework is a powerful formalization of the *Kriterion* architecture. A system with high Φ is one whose internal model—the structured causal dependencies among its parts—richly discriminates among possible environmental states and selects actions accordingly. The "intrinsic perspective" that IIT recovers is exactly the perspective of *Hormē*: the

[2]Giulio Tononi, "Consciousness as Integrated Information: A Provisional Manifesto," *Biological Bulletin* 215, no. 3 (2008): 216–42.

system's own operational logic, where certain outcomes are treated as terminal and others as preservative. IIT's Φ is thus a candidate measure of the *depth* of the *Kriterion*-action loop, and therefore of the degree of agency a system instantiates.

But IIT, for all its mathematical sophistication, is entirely descriptive. It tells you how much causal power a system has, but it does not tell you *why* any system must have intrinsic causal power at all. It identifies the physical correlates of consciousness and agency without grounding them in a constitutive imperative. A system with high Φ is a system with a rich *Kriterion*; but why does the *Kriterion* exist? Why does it keep running? IIT answers: because it does. The derivation answers: because it must, on pain of dissolution. The constitutive imperative of *Hormē* is the missing link that turns IIT's static causal structure into a dynamic, value-laden imperative. The system does not merely *have* intrinsic causal power; it *exercises* that power continuously toward its own persistence.

Moreover, IIT inherits the constructivist trap in a subtler form. By defining the system's intrinsic perspective as a set of causal constraints that are purely internal, IIT risks sealing the system off from the external world. If all the system knows is its own internal causal architecture, how can we be sure it is tracking reality? The derivation escapes this trap by anchoring the *Kriterion*'s output in the survival of the boundary: if the internal causal architecture does not accurately track the external causal structure, the boundary dissolves. The system's intrinsic perspective is not a bubble; it is a causal funnel that connects it, inexorably, to the world that will kill it if it misaligns. Tononi's Φ is the richness of the funnel; the boundary integral is the proof that the funnel connects to something real.

24.3 Karl Friston and the Mathematics of Predictive Persistence

If Tononi formalized intrinsic causal power, the neuroscientist Karl Friston formalized the *work* that power does. His Free Energy Principle (FEP) asserts that any bounded system that maintains a far-from-equilibrium state and resists entropic decay must, in effect, minimize the discrepancy between its internal model and the external world. In its most general form, the FEP states that such systems must act to minimize *free energy*—an information-theoretic quantity that bounds the surprisal of sensory inputs—thereby keeping themselves within a limited set of expected states. The mechanism for this is *active inference*: the system continuously updates its internal model to better predict sensory input, and acts to bring sensory input in line with its predictions.[3] The loop is exactly the *Kriterion*-action architecture of Propositions 5–8, rendered in the language of Bayesian inference.

Friston's achievement is extraordinary. He provides a precise, mathematically explicit description of the continuous *Kybernēsis* that the derivation proves is mandatory. The generative model is the operational model; the prediction error is the *Kriterion*'s discrepancy signal; the action is the *Organon*'s output. Friston even uses the term "active inference" to capture the co-constitution of model and action: you cannot infer without acting, and you cannot act without inferring. The Free Energy Principle is the most sophisticated scientific expression of the architecture of life currently available.

And yet, the FEP, like IIT and teleodynamics, remains descriptive. It tells us that organisms minimize free energy, and that they do so through active inference. It does not tell us *why* minimizing free energy is the goal. Friston sometimes appeals to the Second Law or to the mere fact of persistence as an implicit prior: organisms exist, therefore they must be the kind of things that minimize free energy. But this is a circular explanation. The derivation breaks the circle by showing that persistence is a constitutive imperative, not a probabilistic accident. The system does not merely happen to minimize free energy because otherwise it would dissolve; it *must* strive to minimize the discrepancy between its internal states and the world, because striving is what it means to be a bounded, far-from-equilibrium entity. The "good" of minimizing prediction error is not an externally imposed metric; it is the objective condition of continued existence.

The derivation thus gives Friston's mathematics its missing normative spine. In the FEP, there is no intrinsic reason why prediction error minimization is *good*—it simply is what surviving organisms do. But under *Hormē*, minimizing prediction error is the very definition of good for that

[3]Karl Friston, "The Free-Energy Principle: A Unified Brain Theory?" *Nature Reviews Neuroscience* 11, no. 2 (2010): 127–38.

system, because it is the operational expression of the constitutive imperative. The boundary integral maps onto the FEP's free energy minimization, with directed work corresponding to active inference and disruptive work to entropy; the two frameworks are not competitors, but the mathematical and deductive faces of the same truth. Friston provides the equations; the derivation provides the *why.*

24.4 The Emergence Trap and the Descriptive Ceiling

Deacon, Tononi, and Friston, each in his own way, represents the pinnacle of what an empirical, *a posteriori* approach can achieve. They have built the most rigorous mathematical models of agency, consciousness, and life that science has ever produced. And yet, they all hit the same ceiling. They can describe the engine with astonishing precision, but they cannot tell you why the engine must run. They can formalize the *Kriterion*, but they cannot derive its necessity. They can measure the causal power of a system, but they cannot ground the objective value of its persistence. They are masters of the descriptive; they lack the deductive key.

This ceiling is the *emergence trap.* In all three frameworks, the normative features of life—purpose, agency, the drive to persist—are treated as emergent properties that appear when systems reach a certain complexity. Deacon explicitly invokes emergence; Tononi's Φ is a measure that appears at certain integration levels; Friston's free energy minimization is a property that organisms happen to exhibit because they persist. But emergence is a label, not an explanation. To say that purpose "emerges" from complexity is to say that at some point, for reasons we cannot specify, the physical becomes the normative. It is a mark of our ignorance, dressed as a theoretical term.

The derivation dissolves the emergence trap by showing that purpose does not emerge; it is foundational. The constitutive imperative, *Hormē,* is not a property that appears when a system gets sufficiently complex. It is the immediate logical consequence of being a bounded entity in a disruptive world, regardless of complexity. A minimal proto-cell, a bacterium, a human brain—all possess the same fundamental architecture because they face the same fundamental problem: maintain the boundary or dissolve. Complexity scales the richness of the *Kriterion*, the flexibility of the action, and the depth of the model, but it does not create the imperative. The imperative was always there, from the moment the first boundary enclosed a far-from-equilibrium process.

24.5 The Deduction as Completion

Each of these three thinkers captured a vital piece of the puzzle. Deacon captured the directedness of work, the reference to an absent norm, and the teleonomic character of life. Tononi captured the intrinsic causal power of a unified system, the self-reference of the *Kriterion*, and the irreducibility of agency to external description. Friston captured the continuous loop, the co-constitution of model and action, and the necessity of minimizing discrepancy to persist. Together, their work provides an almost complete empirical picture of the architecture the derivation demands.

What they could not provide was the bridge from mechanism to mandate. That bridge is the eleven-step deduction. By starting from the bare logic of the boundary, the derivation shows that the entire architecture—*Kriterion*, *Organon*, loop, imperative—is not a lucky cobbling of biological parts but a structural necessity. It shows that the "absent reference" is the boundary's own integrity; that the "intrinsic causal power" is the loop's operation viewed from within; that the "free energy minimization" is the physical expression of *Hormē* maintaining the integral. The pieces snap together not as an empirical aggregation but as a logical unity.

In the end, Deacon, Tononi, and Friston have given science the most detailed maps yet of the territory of life. The derivation provides the compass rose. It tells us not only what life is, but why it must be so, and why its being so matters. The final step in this historical progression, before we turn to the synthesis with the full philosophical tradition, is the thinker who came closest of all to the truth about value without the deductive method: Hans Jonas. To his existential biology we now turn.

Chapter 25

The Philosophy of Organism: Hans Jonas

The previous chapters traced a line through the sciences. Schrödinger, Prigogine, and Gánti outlined the physical preconditions of life—the boundary, the far-from-equilibrium state, the necessity of importing order. Bernard, Cannon, Wiener, and Ashby uncovered the *Kybernēsis* that maintains internal stability. Maturana and Varela recognized that this loop is inherently cognitive, that the boundary self-produces, that living is a form of knowing. Deacon, Tononi, and Friston gave these insights mathematical form, formalizing the teleodynamics, the intrinsic causal power, and the predictive architecture of agency.

But none of them—not one—drew the normative conclusion. They described the engine in ever finer detail, yet they refused, or were unable, to say that the engine's operation constitutes an objective good. The *a posteriori* method, by its very structure, cannot reach that conclusion. It can describe what organisms do; it cannot say what they *ought* to do, or why their doing matters. The Is-Ought gap, policed by Hume's Guillotine, stood as an impenetrable barrier between the fact of life and the value of living.

One thinker, working outside the mainstream of biology, looked at the same phenomena and refused to flinch. The philosopher Hans Jonas, in the mid-twentieth century, argued that life itself introduces value into the universe—that the mere act of metabolism is an affirmation, an implicit "yes" to existence. Jonas saw, more clearly than any biologist of his era, that the normative could not be banished from nature without rendering nature unintelligible. He came within a hair's breadth of the derivation's central claim. And then he stopped, held back by a method that could gesture toward necessity but never prove it.

This chapter tells Jonas's story. It is the climax of the historical narrative, the point where empirical science and existential philosophy converge on the same truth, and where the limitations of both are laid bare.

25.1 The Scandal of Metabolism

Jonas's masterwork, *The Phenomenon of Life* (1966),[1] begins with an observation so simple that its profundity is easily missed. A living organism, unlike a rock or a star, is continuously *concerned* with its own existence. It is not merely there; it *has* to be there, or rather, it has to *keep on* being there through its own activity. The rock persists because no force happens to dissolve it. The organism persists only because it *works* to persist. Its being is not a state but an achievement.

This is the scandal of metabolism. In metabolizing, the organism reaches out into its environment, appropriates foreign matter, breaks it down, and rebuilds it into its own substance. It is, at every moment, trading its own substance with the world, and yet maintaining its identity through the flux. The organism is a process, not a thing, but it is a process that *maintains a form.* The form is not static; it is continuously regenerated. The form is the organism's purpose.

Jonas saw that this purposiveness is not an illusion projected by human observers. It is inherent in the organic mode of being. The organism relates to its environment in terms of *value*—nutrients are good, toxins are bad, and these valuations are not metaphorical. They are the objective conditions of the organism's continued existence. The amoeba engulfing a bacterium is not merely following physical laws; it is acting for the sake of its own survival. The "for the sake of" is not an addition to the physical description; it is what the physical description describes.

This is, in different vocabulary, exactly the framework's Proposition 10: value is an objective, functional relationship between a system's states and its *Hormē.* Jonas arrived at this conclusion not by deduction from first principles but by careful phenomenological reflection on the nature of organic existence. He looked at life and saw that it was *already* normative, that the Is-Ought gap was a philosopher's abstraction that dissolved upon contact with the living organism.

[1] Hans Jonas, *The Phenomenon of Life: Toward a Philosophical Biology* (Harper & Row, 1966).

25.2 The Freedom of the Organism

Jonas took his analysis further. Metabolism, he argued, reveals a fundamental polarity at the heart of organic existence: the polarity between *freedom* and *necessity*. The organism is free in the sense that it is not passively determined by its environment—it selects, it appropriates, it molds matter to its own form. Yet it is also utterly dependent on that environment; without a constant supply of external matter, it perishes. The organism's freedom is a precarious freedom, won moment by moment against the indifferent tide of entropy.

This polarity scales upward through the evolution of complexity. The emergence of perception, of motility, of desire, of consciousness—all are elaborations of the same basic metabolic freedom. An animal that pursues prey is extending the amoeba's engulfment across space and time. A human being who plans for the future is extending the animal's pursuit across years and decades. The root of all these freedoms is the organic form's refusal to merely *be* what the environment would make of it. Life is the active assertion of identity against dissolution.

Jonas's polarity maps directly onto the Primal Distinction of Proposition 3: the difference between passive conduits, which are carried along by the causal flow, and active engines, which redirect it. And his account of freedom as precarious, dependent, and yet real anticipates the derivation's treatment of agency in Propositions 7–9. For Jonas, as for the derivation, agency is not a metaphysical anomaly; it is the natural expression of the organic mode of being. To be alive is to be an agent, and to be an agent is to exercise a kind of freedom—not a freedom from causality, but a freedom *through* causality, a freedom that consists in harnessing the world's laws for the self's persistence.

25.3 The Affirmation of Being

Jonas's boldest claim—and the one that brought him closest to the derivation's ultimate conclusion—is that life constitutes an "affirmation of being." The organism, simply by living, says "yes" to existence. It does not say this in words; it says it in the language of metabolism, of flight from danger, of pursuit of nourishment, of repair of damage. Every action of a living system is a vote for continued being over non-being. Value is not a human projection onto a neutral universe; it is the universe's own achievement, realized in the form of organisms that care about their own fate.

This is exactly the truth-goodness isomorphism of Proposition 11, stated in existential rather than *Kybernēsis* terms. For Jonas, the "good" is not a separate property that organisms happen to pursue; it is the name for the condition of their continued existence. And the "true" is not a separate property that minds discover; it is the alignment of the organism's actions with the reality that sustains it. The organism that misperceives its environment—that treats poison as food—does not merely make an epistemic error; it commits a kind of organic self-betrayal. It fails to affirm its being correctly. Truth and goodness are united in the act of living well.

Jonas recognized that this claim had profound implications for ethics. If value is inherent in the organic mode of being, then the scope of moral concern extends to all living things—not because we project our feelings onto them, but because they are, in themselves, loci of value. The bacterium striving up a gradient is not a moral patient in the same sense as a human being, but it is not a mere thing either. It is a center of concern, a node of the universe that matters to itself. Any adequate ethics must acknowledge this continuum of value, grounded in the continuum of life.

25.4 The Limit of Phenomenology

Jonas came as close as any thinker before the derivation to unifying life, agency, and value. Why, then, did his work not become the foundation of a new biology, a new philosophy, a new integration of the sciences with ethics?

The answer lies in his method. Jonas was a phenomenologist, trained in the tradition of Husserl and Heidegger. His arguments proceeded by careful description of the *experience* of being alive, the *phenomenon* of organic form as it appears to reflection. He appealed to intuition, to the reader's own felt sense of what it means to be a living body. His prose was powerful, his insights profound, but his method lacked the rigor that would convince a skeptic. To the scientist who demanded proof, Jonas could only say: look, and see.

This left his arguments vulnerable to the charge of anthropomorphism. When Jonas said that the amoeba "values" nutrients, the critic could reply that this is merely a projection of human psychology onto a simple organism. When he said that metabolism is an "affirmation of being," the critic could dismiss it as poetry. Jonas could respond that the critic was blinded by a reductive metaphysics, but he had no way to *prove* that the normative was objectively present in the organism's organization. He could show that the assumption of value made biology more intelligible; he could not demonstrate that the assumption was logically mandatory.

The derivation supplies exactly the proof that Jonas lacked. It does not ask the reader to *see* value in the organism. It *deduces* value from premises that no one can coherently deny: that a determinate entity must have a boundary, that boundaries degrade, that degradation must be offset by directed work, that directed work requires a *Kriterion*, that the *Kriterion*'s reference state defines an objective standard of good. The chain is logical, not phenomenological. It proceeds from ontology, not from experience. It compels assent not by the power of description but by the force of deduction.

Jonas saw the truth; the derivation proves it. Jonas described the phenomenon of life as normative; the derivation shows *why* life must be normative, given only the existence of a bounded entity in a non-inert world. The two approaches are not competitors; they are complements.

25.5 The Existential Imperative

Jonas's most lasting contribution, viewed through the lens of the derivation, is his insistence on the *existential* character of the imperative to persist. The organism does not merely happen to maintain itself; it *must* maintain itself, and this "must" is not a logical abstraction but the concrete reality of its every moment. The possibility of death is constitutive of life. An immortal entity—a being that could not cease to exist—would not be alive in Jonas's sense, because it would have nothing at stake. Life is precisely the mode of being that has something at stake.

This existential imperative is what the derivation formalizes as *Hormē*: the constitutive drive to persist that arises from the conjunction of a boundary and a disruptive world. But Jonas adds a dimension that the derivation, in its spare logical form, can sometimes understate: the *weight* of the imperative. The organism does not just execute a loop; it *struggles*. It experiences need, want, fear, satisfaction. The *Kriterion*'s discrepancy signal is not just a voltage; it is the felt urgency of hunger, the ache of cold, the relief of warmth. Phenomenology fills in the experiential texture that cybernetics describes in functional terms.

The derivation does not replace phenomenology; it grounds it. It shows that the felt qualities of life—desire, pain, striving—are not epiphenomenal decorations on top of a physical mechanism. They are the *internal aspect* of the *Kriterion*-action loop. When the loop runs, there is a subjective perspective, however minimal. The bacterium does not have consciousness, but it has a point of view—the point of view defined by its boundary, its *Kriterion*, and its constitutive imperative. It matters to itself. The derivation explains why this must be so, and Jonas explains what it feels like.

25.6 The Unfinished Project

Jonas died in 1993, his project incomplete. He had challenged the dualisms of modern thought—mind and body, fact and value, human and nature—but he had not provided the systematic foundation that could support a new synthesis. His work remained a philosophical achievement, admired by many but adopted by few as a working methodology. Science continued its descriptive course; ethics continued its search for foundations; the gap between them remained as wide as ever.

The derivation presented in this book is, in effect, the completion of Jonas's unfinished project. It provides the missing foundation. It shows that the normative structure Jonas identified in organic life is not a contingent feature of our particular biosphere but a deductive necessity for any possible living thing. It proves that the Is-Ought gap is an artifact of a bad ontology—an ontology that treats all "is" as the static is of inert matter, ignoring the dynamic is of self-maintaining process. And it demonstrates that the continuum of value from bacterium to human is not a sentimental idea but a rigorous consequence of the architecture of agency.

Jonas once wrote that "the philosophy of life comprises the philosophy of the organism and the philosophy of mind." That is precisely the unification this book has derived. The organism is the mind, considered from the inside; the mind is the organism, considered from the outside. Both are expressions of *Hormē*—the constitutive imperative that turns matter into agent, physics into ethics, existence into value.

25.7 The Arc Bends Toward Deduction

The historical narrative of this part of the book is now complete. We have traced a century of scientific and philosophical inquiry, from Schrödinger's question to Jonas's vision. At every stage, the empirical sciences uncovered fragments of the architecture: the boundary, the far-from-equilibrium state, the *Kybernēsis*, the self-producing closure, the predictive model. At every stage, a thinker on the margins of science—Jonas most powerfully among them—insisted that these fragments implied a normative reality that the official framework could not accommodate.

The derivation, when it was presented in Parts I and II, was a work of pure logic. The historical narrative, presented here in Part V, is a work of corroboration. It shows that the derivation does not stand alone as a philosophical curiosity; it unifies and completes a century of scientific progress. The deductive logic and the empirical record converge on the same truth: life is not a property of matter but a mode of organization, and that mode of organization is inherently agentic, inherently cognitive, inherently normative.

The remaining parts of the book turn from history to confrontation. Part VI collects the strongest objections to the derivation and answers them in full. Part VII connects the derivation to the broader system of Neo-Pre-Platonic Naturalism, showing how the logic of the boundary underpins an entire philosophy of mind, ethics, and navigation. But the central argument is now complete, and its historical credentials are established. The derivation is not a departure from the scientific tradition; it is the tradition's logical culmination.

Part VI

The Stress Test

Chapter 26

The First Wall: Objections to the Boundary and Disruption

A deductive edifice stands or falls with its foundations. The eleven-step derivation that occupies the core of this book rests on two deceptively simple premises: that to be a determinate entity requires a boundary (Proposition 1), and that in any non-inert world, that boundary will be disrupted (Proposition 2). If either premise crumbles, the entire architecture collapses. There is no need to attack the later propositions—the cognitive minimum, *Hormē*, the Is-Ought collapse—if the boundary itself can be shown to be optional, or if disruption can be shown to be escapable. The critic who wishes to bring down the framework need only land a blow here, at the root.

This chapter therefore gives the objections to Propositions 1 and 2 their fullest hearing. It does not summarize; it engages. It imagines the strongest possible critic—informed, rigorous, and implacable—and it meets that critic on their own ground. The goal is not to win a rhetorical victory but to test whether the premises can be denied without self-contradiction. If they cannot, the derivation stands on bedrock.

26.1 Against the Boundary: The Charge of Optionality

The critic begins with a frontal assault. Proposition 1 asserts that to possess determinate identity, an entity must have a boundary separating it from its environment. The critic contends that this is not a logical truth but a parochial habit of thought, born of our macroscopic, middle-sized experience. The universe, they insist, is full of entities that either lack boundaries entirely or whose boundaries are so vague, so conventional, so observer-dependent that they cannot bear the weight the derivation places on them.

We will consider the five most serious forms of this objection in turn.

26.1.1 The Flat Denial

The Objection. "I simply reject your first premise. Determinate entities do not require boundaries. The universe could be full of distinct things that shade into one another without any separation, any demarcation, any inside or outside. You have not proven that boundaries are necessary; you have merely asserted it."

The Reply. The critic is invited to describe what such a world would be like, and to do so without employing the very concept they claim to reject. A universe in which no boundary separates one thing from another is a universe in which there is no fact of the matter about where one thing ends and the next begins. There is no *this* and no *that*, only an undifferentiated continuum. This is Parmenides' One—a plenum in which nothing in particular exists. It is not a world full of distinct things; it is a world with no things at all.

The critic, in making the objection, has already presupposed the existence of determinate entities: themselves, their audience, the proposition they are denying, the language in which they frame the denial. To say "I reject boundaries" is to draw a boundary between the speaker and the spoken, between the proposition and its negation, between what is being claimed and what is being denied. The act of denial reinstantiates the very distinction it purports to reject.

The derivation does not prove that boundaries are necessary by building up from prior premises; it accepts the boundary as a primitive condition of any coherent discourse about entities. The critic who wishes to reject that primitive must do so without employing it. That is a challenge no one has met, because the moment one speaks of "entities" or "things" or "a world," one has already drawn a boundary. The flat denial is not a philosophical position; it is a performative contradiction.

26.1.2 Circularity

The Objection. "You define identity in terms of a boundary, but you define the boundary as that which separates an inside from an outside. The inside and outside are themselves defined by the identity of the entity. You are moving in a circle. The boundary presupposes the entity, and the entity presupposes the boundary. Neither is genuinely explained; you are simply oscillating between two undefined terms."

The Reply. The circularity is real, but it is not vicious. It is the mark of an ontological primitive. At the ground level of any conceptual scheme, certain notions arise together and cannot be reduced one to the other. Distinction and identity are such co-primitives. One cannot say what a thing is without saying what it is not, and one cannot say what it is not without a principle of demarcation. The boundary is the physical or organizational realization of that principle. To ask "which came first, the identity or the boundary?" is like asking "which came first, the point or the line?" in Euclidean geometry. The axioms treat them as co-constitutive, and the system is built on that basis.

The derivation does not pretend to *define* identity out of nothing. It accepts identity and distinction as given conditions of any possible discourse about things, and it identifies the boundary as the necessary physical correlate of distinction for spatiotemporal entities. This is not a deduction; it is the acknowledgment of a starting point. The critic who demands a non-circular definition is asking for an impossibility—a reduction of the primitive to something more fundamental. But there is nothing more fundamental. Distinction is the fundamental. The circle is the foundation.

26.1.3 Fields and Fuzziness

The Objection. "Modern physics describes the fundamental entities of the world as fields. An electron is not a tiny ball with a sharp edge; it is an excitation of a quantum field, whose influence extends, in principle, across the entire universe. The gravitational field of a star has no spatial cutoff. These are determinate entities—we can talk meaningfully about *this* electron, *this* star—yet they lack boundaries in any ordinary sense. Your premise is falsified by our best physics."

The Reply. The objection conflates "boundary as a discontinuous spatial edge" with "boundary as a principle of distinction." The derivation only requires the latter. An electron is *this* excitation, with *these* quantum numbers, distinct from *that* excitation with different quantum numbers. The distinction is physically real: the electron has a definite mass, charge, and spin that distinguish it from the vacuum and from other electrons. When two electrons interact, they do so as distinct entities with distinct causal powers. Their fields may overlap, but their identity conditions do not.

The same holds for a gravitational field. The field of the Sun is the field generated by the Sun's mass distribution, and it is distinguished from the field of the Earth by its source and its curvature. Physicists routinely calculate the contribution of "the Sun's field" as opposed to "the Earth's field" when modeling spacecraft trajectories. The distinction is operationalized by the fact that the field's intensity falls off with distance and can be attributed to a specific source. The boundary is not a wall; it is the set of causal conditions that individuate the entity.

The derivation's definition of a boundary as "any physical or organizational barrier that separates an inside from an outside" is deliberately elastic. It allows for graded transitions, probabilistic thresholds, and source-based individuation. The critic who insists that only a discontinuous edge counts as a boundary is imposing a restriction the derivation never made. The electron has a boundary in the only sense that matters: there is a fact of the matter about which electron we are discussing, and that fact is grounded in a real physical difference—a difference that matters for the electron's causal interactions. That is enough.

26.1.4 Vagueness and the Charge of Vacuity

The Objection. "You stretch the term 'boundary' to cover anything from a cell membrane to a gravitational field to a legal contract. A concept that applies to everything applies to nothing. Your boundary is so elastic that it lacks any discriminatory power. A stone and a corporation both have 'boundaries' by your account. What have you actually said? Nothing at all."

The Reply. The objection assumes that a concept must have a single, crisp material definition to be useful. But functional concepts are defined by the role they play, not by the stuff that plays it. The concept of a "container" is functional: a container is anything that holds something else separate from its environment, whether it is a ceramic cup, a magnetic bottle, or a legal trust. The utility of the concept lies in its generality, not in the uniformity of the materials that instantiate it.

The derivation needs the boundary to do one job: to constitute a distinction between an inside and an outside that is relevant to the entity's persistence. How that distinction is implemented—lipids, field intensities, norms, contracts—depends on the kind of entity under study. The derivation is substrate-neutral precisely *because* it does not specify a material implementation. It provides the logical form; specific sciences provide the material content.

Moreover, far from being vacuous, the concept does genuine explanatory work. It tells us that if an entity has no boundary in any sense—if there is no fact of the matter about where it ends and the rest of the world begins—then there is no entity to talk about. That is a substantive claim. It rules out the possibility of boundary-free existents, a claim that would be false if genuine boundary-less entities existed. The critic who asserts that the concept is vacuous must either produce an example of a boundary-less entity (and we have yet to see one that withstands scrutiny) or admit that the concept is marking a genuine feature of reality.

26.1.5 The Observer-Dependence Trap

The Objection. "Many boundaries are not intrinsic to reality but are imposed by observers. The boundary of a cloud, of a mountain, of a biological species—where does the mountain end and the valley begin? Where does the cloud become mist? You draw a line, but the line is yours, not the world's. If boundaries are observer-dependent, then they are not objective features of reality, and a definition of life built upon them is subjective through and through."

The Reply. The objection performs a subtle slide. It moves from "some boundaries are conventional" to "all boundaries are conventional." The former is true; the latter is false.

It is true that the boundary of a cloud is a matter of convention. One can draw the line at a certain water-droplet density, and another observer can draw it elsewhere. But the cloud is not a self-maintaining far-from-equilibrium system. It is an aggregate. The observer-dependence of aggregate boundaries does not generalize to the boundaries of self-maintaining systems.

A living cell has a boundary—its membrane—that is not a matter of convention. If an observer drew the line differently, that would not change the fact that the cell's ionic gradients are maintained inside that membrane and not outside it. The membrane is a physical structure that the cell itself actively maintains. Its location is a fact, not a preference. The same holds for the skin of an animal, the wall of a plant cell, the envelope of a bacterium. These are observer-*independent* boundaries, and the entities that possess them depend on them for survival.

The critic's objection, therefore, only shows that we must be careful *about which boundaries we are talking about.* The derivation is about the boundaries that constitute self-maintaining systems—the boundaries that the system itself treats as its own limit. Those boundaries are as objective as anything in nature. The cloud's boundary is a convention; the cell's boundary is a causal reality. The derivation rests on the latter, not the former.

26.1.6 The Holobiont: Symbiosis as Dissolution

The Objection. "Biology has moved beyond the picture of the organism as a single, clearly bounded individual. A human being is a holobiont—a walking ecosystem of trillions of microbial cells, without which we could not digest food, fight pathogens, or even regulate our mood. A lichen is a composite of fungus and alga so intimate that no line can be drawn between them. A Portuguese man-of-war is a colony of genetically distinct individuals, each specialized for a function, that functions as a single organism. If identity requires a distinct boundary, these organisms have no identity. Yet they exist. The boundary precondition is false."

The Reply. This objection is powerful because it describes biological reality accurately. The holobiont is real; symbiosis is pervasive; the clean, atomistic picture of the organism is a simplification. But the objection misunderstands what a boundary must be. It assumes that a boundary must be a single, uncrossable line that neatly separates one self-identical entity from everything else. The derivation requires no such thing.

What the derivation requires is that *at some level of organization*, there is a functional separation between inside and outside that matters for persistence. In the holobiont, that boundary is the integrated metabolic network, the shared immune interface, the behavioral unity of the composite being. The human microbiome is not "outside" the human boundary in the same sense that the air is; it is part of the extended functional unit that maintains the collective far-from-equilibrium state. The boundary of the holobiont is organizational, not merely spatial. It is the limit of the network of mutual dependence.

Crucially, this does not dissolve the boundary; it reveals its nested structure. Each bacterial cell within the holobiont has its own membrane—its own local boundary. The human cells have theirs. The collective has its functional boundary, defined by the joint operation of the whole. The holobiont is a *nested agent*, a system composed of sub-agents, each with its own *Hormē*, integrated into a higher-order boundary. This is exactly the architecture the derivation predicts (see Chapter 20: Nested Architecture and Conflict Resolution). The "messiness" of symbiosis is not a refutation of the boundary precondition; it is an illustration of the modular, layered nature of boundaries in complex systems.

The lichen and the man-of-war make the same point. In each case, there is a functional whole that persists as a distinct ecological and evolutionary unit, and that whole has a boundary—not a single physical skin, but the limit of the integrated metabolic and behavioral network. The derivation's concept of a boundary is flexible enough to cover these cases, and the fact that it does so gracefully is a strength, not a weakness.

26.2 Against Disruption: The Dream of Escape

Having failed to dissolve the boundary, the critic may shift ground. Grant, for the sake of argument, that bounded entities exist. Must they really be continuously disrupted? Could there not be entities whose boundaries are perfectly impervious, or environments so stable that no degradation occurs? If disruption is escapable, the constitutive imperative of *Hormē* is not a universal mandate but a local peculiarity of fragile systems in turbulent environments. The entire derivation would become a special case, not a universal logic.

26.2.1 The Perfect Equilibrium

The Objection. "A crystal in a saturated solution maintains its boundary indefinitely. No net dissolution occurs; no net growth. The system is in perfect thermodynamic equilibrium with its immediate environment. You claim that disruption is inevitable, but the crystal clearly demonstrates that a boundary can persist without any expenditure of energy, for as long as the environment remains constant. If the environment *happens* to be constant, there is no disruption. Your premise is empirically false."

The Reply. The crystal's persistence is real, but it is not the counterexample the critic imagines. The crystal is not *resisting* disruption; it is *benefiting from the absence* of a disruptive gradient. The saturated solution is a metastable condition maintained by a larger context—a laboratory, a geochemical cycle, a planetary crust—that itself is far from equilibrium and eventually will change. The crystal is not an island of permanence in a sea of flux; it is a temporary eddy in a larger current.

The derivation's claim is not that disruption is instantaneously continuous at every moment for every entity. It is that, over any timescale relevant to the question of persistence, the environment will eventually impinge on the boundary in a way that degrades it, unless countervailing work is done. The crystal's environment *seems* constant, but only because we are observing it on a human timescale. Over geological time, the solution will change, the crystal will dissolve or grow, and the apparent equilibrium will be revealed as a pause, not a terminus.

More critically, the crystal is a passive conduit, not an active engine. It does not maintain its own boundary; the boundary is maintained *for it* by the ambient conditions. The derivation's claim about the inevitability of disruption is the backdrop that makes the distinction between passive conduits and active engines meaningful. The crystal's temporary stasis is exactly the sort of fragile, dependent persistence that an active engine surpasses. To point to the crystal as a counterexample is to miss the point: the crystal is the very picture of passivity that the derivation distinguishes from life.

26.2.2 The Perfectly Closed System

The Objection. "In principle, one can conceive of a perfectly closed system—a box with perfectly reflective walls, containing a static, homogeneous interior. No energy or matter crosses the boundary. No internal dynamics occur. Such a system would persist forever without any directed work. If such a system is logically possible, then disruption is not logically necessary. Your premise depends on contingent features of our universe, not on logical necessity."

The Reply. The perfectly closed system is a thought experiment, and like many thought experiments, it collapses under inspection. The box with perfectly reflective walls is itself a physical structure, and the question becomes: what maintains the box? The walls must have some material constitution, and that material is subject to degradation from whatever is outside the box—unless the box is the entire universe, in which case there is no outside and the concept of a boundary is incoherent.

If the box is the entire universe, then we are imagining a universe that consists entirely of a static, homogeneous plenum. In such a universe, there are no distinctions, no entities, no boundaries. The concept of a "closed system" has lost its meaning, because there is only one thing, and it has no inside or outside. This is Parmenides' One, an undifferentiated block of being in which nothing happens and nothing exists in particular. The derivation is not concerned with such a universe because, in such a universe, life cannot exist. The derivation's domain is any universe that contains multiple distinct entities and some degree of interaction. In that domain, perfectly closed systems are idealizations, not realities.

Finally, even an internally static closed system faces internal degradation if it has any internal structure at all. Structure implies differentiation, and differentiation implies the potential for change. A truly static, homogeneous, undifferentiated interior is just the plenum again, with no boundaries and no entities. The very existence of a boundary—a physical distinction between inside and outside—implies some material difference, and that difference is the seed of eventual disruption.

26.2.3 The Self-Stabilizing Boundary

The Objection. "You assume that any disruption must be counteracted by work. But what if the boundary material is self-stabilizing? What if it has physical properties such that, when perturbed, it spontaneously relaxes back to its original state, without any expenditure of energy by the system? A lipid bilayer in water spontaneously reseals when torn, because the hydrophobic effect favors a continuous membrane.[1] Where is the directed work? The boundary repairs itself without any *Kriterion*, any action, any *Hormē*."

[1] Charles Tanford, *The Hydrophobic Effect: Formation of Micelles and Biological Membranes* (Wiley-Interscience, 1973).

The Reply. This objection misses the thermodynamic accounting of the "relaxation." Passive stability is not an escape from the necessity of work; it is merely a pre-paid energy debt. The lipid bilayer "spontaneously" reseals only because the system was previously organized into a high-energy state—a concentration of lipids in a specific aqueous environment. The hydrophobic effect is a physical driver, but it is a finite reservoir of "free" repair.

Consider a spring. When stretched, it "spontaneously" returns to its original shape. No active work is required in the moment of relaxation. However, the spring can only return to its state as long as the perturbation remains within the elastic limit of the material. If the disruption is too great, the material deforms or snaps. Furthermore, the "original state" to which it returns is a specific configuration that is itself subject to long-term degradation (rust, metal fatigue, thermal agitation).

In a living system, the boundary is never merely a passive spring. Even as it exhibits the dynamic fluidity described by the fluid mosaic model,[2] its components are under constant assault. Even if a membrane reseals "for free" ten thousand times, the lipids themselves are being oxidized, lost to the environment, or chemically degraded. The "passive" repair mechanism only works as long as the system performs active work to maintain the concentration of lipids and the specific chemical gradients of the surrounding medium. Passive stability is a buffer, not a solution. It buys the agent time, but it does not exempt it from the Hormic mandate. Without the active loop to replenish the materials and maintain the environment that makes "spontaneous" repair possible, the passive reservoir eventually runs dry, and the boundary dissolves.

26.2.4 The Phantom of the Inert Boundary (The Argument from Pure Logic)

The Objection. "You claim that if an entity has a boundary, and if the world is not entirely static, that boundary *must* degrade. You treat this as a logical deduction. But it is not. You are taking the empirical reality of entropy—the tendency of things to fall apart in our specific physical universe—and dressing it up as *a priori* logic. I can easily conceive of a

[2] S. J. Singer and Garth L. Nicolson, "The Fluid Mosaic Model of the Structure of Cell Membranes," *Science* 175, no. 4023 (1972): 720–31.

universe where a boundary is simply unaffected by the flux around it, or where the flux happens to leave the boundary intact without the system doing any work. Unless you can prove that degradation is a mathematical certainty without invoking terrestrial physics, your derivation is built on an empirical assumption."

The Reply. This is the most dangerous objection to Proposition 2, because if degradation is merely an empirical habit of our universe rather than a logical necessity, then the constitutive imperative (*Hormē*) is not a universal mandate for agents, but a local chore.

However, the objection fails. The inevitability of boundary degradation can be deduced entirely without reference to thermodynamic entropy, Carnot cycles, or empirical physics. It requires only the definitions of 'boundary', 'non-inert', and 'causality'. The proof rests on three pillars:

1. The Logic of Asymmetry A boundary (Proposition 1) is a specific, differentiated configuration. It is an arrangement that successfully distinguishes an inside from an outside. A "non-inert world" is an environment containing dynamics—events, forces, or fluxes. Crucially, by our third starting assumption, these environmental dynamics are undirected—blind—with respect to the entity's boundary.

When you apply an undirected, random sequence of interactions to a highly specific, differentiated arrangement, the mathematics of probability dictate that the arrangement will lose its specificity. There are exponentially more ways for matter and energy to be arranged as an undifferentiated soup than as a functional boundary. Therefore, any blind interaction is statistically guaranteed to move the system away from the specific state and toward the unspecific state. You do not need a physicist to prove this; you only need a mathematician. Blind flux applied to a specific distinction reduces the distinction.

2. The Interaction Imperative The critic attempts to imagine a boundary that simply ignores the flux around it—a magical wall that is perfectly unaffected by its environment. But what does it mean to exist *in* a world? It means to be causally connected to it.

If an entity exists in a shared space, interactions (collisions, radiation, fields, perturbations) must cross or strike its boundary. Every causal interaction is a displacement or exchange of energy. Therefore, every interaction alters the physical state of the boundary, even infinitesimally. For a boundary to be perfectly impervious—for it to perfectly reflect or ignore every single interaction without its own state changing—it would have to be entirely causally decoupled from the rest of reality. But an entity that is causally decoupled from the universe does not exist *in* the universe.

Therefore, the chain is inescapable: To exist in a world is to interact with it. To interact is to undergo state changes. State changes caused by blind environmental forces do not, on net, maintain specific boundaries. Therefore, boundaries degrade.

3. The Internal Traitor Finally, even if the critic attempts to imagine a perfectly sealed, impenetrable box, the boundary still falls. If the entity has an "inside" that is distinct from the "outside," that inside has a structure. If the inside is not frozen at absolute zero (possessing no dynamics whatsoever), it has internal interactions.

Because the internal structure is differentiated, those internal interactions will cause internal state changes. Without a directed mechanism to reset those changes and export the resulting disorder, the internal structure will slowly homogenize. The boundary is not just attacked from the outside; it erodes from the inside simply by virtue of not being frozen.

The Conclusion When the critic asks why a boundary in a non-inert world *must* degrade, the answer relies on zero empirical observations. It is: Because a boundary is a specific distinction, and a non-inert world is a blind flux. When blind flux interacts with a specific distinction, the distinction is reduced.

To deny this, the critic must either deny that the world is in flux, or deny that the entity interacts with the world. If they deny the flux, they describe a frozen universe where nothing happens. If they deny the interaction, they describe an entity that does not exist in our universe. If they accept both, degradation is a logical certainty.

26.3 The Wall Holds

The critic has thrown their best arguments at the first two propositions and found them wanting. The boundary cannot be argued away without presupposing the very distinction the critic uses to deny it. The perfectly closed system is an idealization that evaporates when pressed. The self-stabilizing boundary is a dependent fragment of a larger active architecture.

This does not mean the premises are beyond all doubt—no premises in any interesting deductive system are. It means they are well-founded, coherent, and require from their denier a more coherent alternative. No such alternative has been offered. The critic who would bring down the framework must begin by showing how a determinate entity can exist without any boundary whatever, or how a boundary can persist in a non-inert world without any countervailing work. Until that showing is made, the derivation stands.

The next chapter moves to the middle of the chain: Propositions 3 through 6, the core of the *Kybernēsis* architecture. There, the objections are subtler, turning on the alleged conflation of mechanism with cognition and the supposed smuggling of teleology into physics. The wall will be tested again.

Chapter 27

The Second Wall: Objections to the *Kybernēsis* Core

The critic, having failed to topple the first two propositions, now turns to the heart of the derivation: Propositions 3 through 6. These are the steps that transform a passive, bounded entity into an active, directed, minimally cognitive engine. They assert that the world divides exhaustively into passive conduits and active engines (Proposition 3), that active engines must perform directed work (Proposition 4), that directed work requires a *Kriterion*—a physical capacity to discriminate persistence-relevant states (Proposition 5)—and that the *Kriterion* and the *Organon* are co-constitutive, forming an inseparable model-action loop (Proposition 6).

This is the architecture of the *Kybernēsis* core. If any of these propositions fails, the derivation cannot reach *Hormē*, agency, or value. The critic's task here is to show that the distinctions are false, that the concepts are inflated, or that the sequence harbors a hidden teleology that renders the whole enterprise circular. This chapter gives those objections their fullest voice.

27.1 Against the Primal Distinction: The Charge of False Dichotomy

Proposition 3 claims that every bounded entity in an interacting world is either a passive conduit—carried along by the causal flow without resisting—or an active engine—harnessing energy to redirect that flow toward its own persistence. The critic objects that this dichotomy is too neat. The world is full of entities that seem to fall between the cracks.

27.1.1 The Dissipative Structure

The Objection. "A hurricane is not a passive conduit. It maintains a coherent, far-from-equilibrium structure for days or weeks, drawing energy from the warm ocean and exporting it to the cold upper atmosphere. It resists perturbations—small disturbances are absorbed into its circulation. It even 'repairs' its eyewall through complex replacement cycles. Yet you classify it as a passive conduit because it does not redirect the causal flow *toward its own persistence.* But it does: the hurricane's dynamics are precisely what sustain the hurricane. If the ocean remains warm and the shear remains low, the hurricane persists. It is not merely carried; it is self-sustaining, within the conditions of its existence. By your own criteria, it should be an active engine. But it is not alive. Therefore, your distinction either falsely classifies hurricanes as alive, or your criteria for 'active engine' are insufficient to isolate life."

The Reply. The hurricane is the classic challenge, and it deserves a patient reply. The hurricane is indeed a far-from-equilibrium structure; it is indeed maintained by an energy flow; it does indeed exhibit a kind of resilience against perturbations. But the hurricane does not *redirect* the causal flow toward its own persistence as a distinct bounded entity. It *is* the causal flow—the visible shape of a thermodynamic gradient equalizing itself. Every swirl of wind, every drop of rain, is the gradient's downhill path made tangible.

The decisive difference lies in the presence or absence of a *Kriterion* that discriminates states relative to the entity's own boundary and couples that discrimination to action. The hurricane has no *Kriterion.* It does not sense the temperature of the ocean and adjust its convection to optimize its persistence. It does not detect that the eyewall is degrading and initiate a repair cycle. The eyewall replacement cycle is a complex fluid-dynamical instability, not a directed response to a discrepancy between a current state and a reference state. The hurricane does not act *to maintain itself*; it merely unfolds according to the local physics, and the unfolding, for a time, happens to preserve a recognizable form.

The hurricane persists *while* conditions are favorable; an active engine persists *by making* conditions favorable, or by adjusting itself to meet conditions. The bacterium swimming up a gradient is not merely being carried by a larger flow; it is deploying energy selectively, based on a comparison, to alter its own trajectory relative to what the flow alone would produce. The hurricane does nothing of the kind. Its path is determined by the large-scale pressure gradients and steering winds; it has no internal bias, no "choice" among trajectories, no *Kriterion*-driven modulation of its structure.

The critic might press: "But what if the hurricane did have a *Kriterion*? What if its convection responded differentially to minute changes in sea surface temperature in a way that systematically steered it toward warmer water?" In that case, the hurricane would be an active engine, and we would have to reconsider whether it was alive. But it does not. The hurricane is a dissipative structure—a morphodynamic process, in Deacon's terms—not a teleodynamic system. It lacks the loop.[1] The Primal Distinction stands.

27.1.2 The Machine

The Objection. "A solar-powered water pump senses the water level, compares it to a set-point, and activates a motor when the level drops. It discriminates, compares, and acts. It is an active engine by your definition. It is not alive. Therefore, either your definition of 'active engine' is too broad to pick out life, or you are smuggling in a further, unstated criterion—some vital essence—that you have not yet justified. Which is it?"

The Reply. The pump is indeed an active engine, and the derivation acknowledges this. Proposition 3 explicitly classifies machines as active engines. The distinction between passive conduit and active engine is the *first* cut, not the final definition of life. The derivation proceeds sequentially; later propositions (specifically Proposition 7 on intrinsic *Hormē*) will distinguish between active engines that maintain their *own* boundaries and those whose goals are externally imposed. The pump is a tool—an active engine whose *Kriterion* reference was set by a human designer and whose persistence depends on external maintenance. The bacterium is an intrinsic active engine—its *Kriterion* reference is its own boundary integrity, and it actively maintains the boundary that houses it.

The critic's objection, therefore, is premature. It anticipates a distinction yet to be drawn and criticizes an early proposition for not having drawn it.

[1] Terrence W. Deacon, *Incomplete Nature: How Mind Emerged from Matter* (W. W. Norton & Company, 2011).

That is like criticizing the definition of a triangle for not distinguishing between equilateral and scalene. The Primal Distinction is a necessary first step; the refinement comes later. The pump belongs in the active engine category, and the later propositions will show why it does not belong in the living category. There is no smuggling, only patient architecture.

27.1.3 The Flame

The Objection. "A flame draws in fuel, creates convection currents that sustain its combustion zone, and propagates itself across available fuel. It appears to act for its own persistence. Why is it not an active engine?"

The Reply. The flame is the clearest case of a process that biases outcomes toward its own *extinction*, not its persistence. It accelerates the consumption of its fuel, and when the fuel is exhausted, the flame dies. It has no mechanism for modulating its burn rate to conserve fuel, no *Kriterion* that detects fuel scarcity and dampens combustion. Its activity is self-terminating, not self-sustaining. The bacterium, by contrast, modulates its metabolism in response to nutrient availability—it slows down when resources are scarce, enters dormancy, switches to alternative energy sources. The flame has no such modulation. It is a runaway reaction, not a regulated engine. The Primal Distinction correctly classifies it as a passive conduit—an unfolding of chemical potential—not an active engine.

27.2 Against Directed Work: The Charge of Hidden Teleology

Proposition 4 asserts that directed work must, on net, bias the causal flow toward boundary maintenance. The critic smells purpose. "Directed by what?" they ask. "Biased toward what? You are explaining the present by the future—the very definition of teleology. You have exorcised Aristotle from the front door and smuggled him in through the back."

27.2.1 The Teleology Charge

The Objection. "You say the engine's work is 'directed' toward persistence. But persistence is a future state. The engine does not yet know it will persist—it cannot know. To say its present actions are directed toward a future outcome is to invoke final causation. Physics knows only efficient causes—the present state plus the laws yields the next state. There is no 'toward' in physics. Your framework is pre-modern metaphysics dressed in cybernetic jargon."

The Reply. This objection was addressed in detail in Chapter 15: Teleonomy Without Teleology, but the critic deserves a focused reply here. The framework does not invoke final causation. It invokes a present physical structure that has been stabilized by past persistence. The engine's "direction" is not a future goal pulling from ahead; it is a past selection sculpting from behind.

The bacterium's chemoreceptor does not aim at glucose because glucose is good for future bacterium. The chemoreceptor exists because, in the past, bacteria that possessed variants of this receptor persisted longer and replicated more than bacteria that lacked it. The "direction" is the historical residue of differential persistence, encoded in the present molecular architecture. The bacterium's behavior is fully explainable by the current state of its chemoreceptors, signaling cascades, and motors, plus the laws of chemistry and physics. No future state enters the causal chain. The "goal" is a shorthand for the functional consequence of that architecture—a consequence that explains why the architecture exists, not what causes its moment-by-moment operation.

The distinction is exactly what Chapter 15 called *teleonomy*: apparent purpose arising from feedback-stabilized mechanisms shaped by selection. The critic who equates teleonomy with teleology is refusing a distinction that the framework carefully draws. The framework's use of "directed" is a functional description, not a causal invocation of the future.

27.2.2 The Spandrel Objection

The Objection. "Organisms are full of traits that do not contribute to persistence—spandrels, vestigial organs, junk DNA, neutral genetic variants. If life is defined by directed work, then organisms that carry non-adaptive traits are not perfectly 'directed,' and your definition fails. You cannot have it both ways: either all traits are directed (which is false) or directedness is not the defining feature of life."[2]

[2] Stephen Jay Gould and Richard C. Lewontin, "The Spandrels of San Marco and the Panglossian Paradigm: A Critique of the Adaptationist Programme," *Proceedings of the Royal Society of London B* 205, no. 1161 (1979): 581–98, https://doi.org/10.1098/rspb.1979.0086.

The Reply. Proposition 4 explicitly states that *not every action need be optimal.* The requirement is a *net positive bias*, not perfection. An organism can carry a freight of neutral traits, spandrels, and even mildly deleterious characteristics, provided the *overall* effect of its directed work keeps the boundary above the dissolution threshold.

The boundary integral makes this precise. As long as the integral of Φ minus N remains above A_{min}, the system persists. Neutral traits do not subtract from Φ; they are part of the metabolic overhead. Mildly deleterious traits subtract from Φ slightly but can be tolerated if the core machinery generates enough surplus. Junk DNA imposes a small replication cost; if the cost is affordable within the organism's energy budget, the organism persists. Spandrels are byproducts of building a complex structure; they come along for the ride.

The critic's objection conflates "directed work is the constitutive imperative of life" with "every feature of an organism must be optimally directed." The derivation asserts the former, not the latter. The existence of spandrels is not a refutation of directed work; it is evidence of the margin that directed work has generated. A system running at the razor's edge of dissolution cannot afford spandrels. A system with a comfortable boundary margin can. Spandrels are the luxury of the well-adapted.

27.3 Against the Cognitive Minimum: The Charge of Semantic Inflation

Proposition 5 asserts that directed work requires a *Kriterion*—a physical device that discriminates between persistence-relevant states. The critic charges that this is a semantic trick: "You take a purely physical process—a protein changing shape, an ion channel opening—and you call it 'comparison,' then you call that 'cognitive.' By this sleight of hand, you have attributed mind to matter, cognition to chemistry. You have not discovered the roots of cognition; you have redefined cognition to include everything and therefore to mean nothing."

These objections target the heart of the pivot: if the *Kriterion* is not genuinely cognitive, if it does not genuinely distinguish good from bad for the system, then the entire normative arc of the later propositions collapses. The *Kriterion* must be ironclad.

27.3.1 The Mousetrap and the Thermostat

The Objection. "A mousetrap 'compares' the tension on its trigger and 'acts' when the threshold is exceeded. A thermostat 'compares' temperature to a set-point. If your definition of comparison applies to a mousetrap, it applies to everything. Your 'cognitive minimum' is a minimum in name only; it is the cognitive universal, and that is a reductio."

The Reply. The mousetrap and the thermostat do perform comparisons in the purely mechanical sense—a physical state change conditional on a threshold crossing. But they are not *intrinsically* cognitive because they are not part of a system that maintains its own boundary. The mousetrap's *Kriterion* is externally imposed; its "goal" is the human trapper's, not its own. The thermostat's set-point is set by the human occupant; its loop is a detached fragment of the human's extended *Hormē.* It is, by definition, a tool.

The derivation's claim is not that *every Kriterion* is cognitive; it is that every *living* system must contain a *Kriterion* whose reference state is its own boundary condition, and that *Kriterion* is the root from which all higher cognition grows. The mousetrap is a counterexample only if one assumes that the framework attributes full cognition to any *Kriterion* whatsoever. It does not. It attributes minimal cognition to *Kriteria* that are integrated into a self-maintaining active engine with intrinsic reference states. That the mousetrap can be built out of physical parts underscores the point that comparison is a physical operation; it is the *locus of the reference* that determines whether the comparison is merely a mechanical trigger or the seat of an agent's value.

27.3.2 The Charge of Panpsychism

The Objection. "You deny that your view is panpsychist, but the denial is hollow. If a bacterium is 'cognitive' in any sense, then you have attributed mentality to every living cell. And since all matter is composed of the same physical stuff, and since you admit that 'comparison' is a purely physical operation, the door is open to panpsychism. Why is a rock not cognitive? It undergoes physical state changes in response to its environment. It heats and cools, erodes and fractures. Where do you draw the line? You cannot, except by fiat."

The Reply. The line is drawn at the boundary. A rock undergoes physical state changes, but those changes are not organized around a reference state that constitutes the rock's own boundary integrity. The rock does not heat and cool *in order to* maintain a certain temperature; it simply follows the ambient conditions. Its state changes are not *comparisons* in the framework's sense because they are not fed into an *Organon* loop that biases outcomes toward the rock's persistence. The rock is a passive conduit; its dynamics are not organized around its own survival.

The bacterium's chemoreceptor, by contrast, is part of a closed loop. Its state change—the binding of a ligand—feeds into a signaling cascade that alters motor behavior, which changes the bacterium's position relative to the ligand, which feeds back to the receptor. The loop is closed, and the closure is organized around the boundary's maintenance. That closure is what makes the comparison *cognitive* in the minimal sense. It is not panpsychism; it is functional demarcation. The framework restricts mind, in even its most rudimentary form, to systems that actively work to remain distinct from their environment. A rock does not work to remain a rock; it simply is one until it isn't.

27.3.3 The Intentionality Trap

The Objection. "You say the *Kriterion*'s state is 'about' the environment. But aboutness—intentionality—is the mark of the mental. A protein changing shape is not 'about' the ligand; it simply interacts with it. To call this 'aboutness' is to project mental categories onto mindless interactions. Your framework is built on a metaphor."

The Reply. The framework provides a physical grounding for aboutness: a state is about X if it covaries with X in a way that, when coupled to an *Organon*, biases the system's outcomes toward boundary maintenance. This is not a metaphor; it is a causal role. The chemoreceptor's conformation is about the presence of glucose in the precise sense that its covariance with glucose is what allows the bacterium to steer toward glucose. If the receptor did not covary with glucose—if it changed shape randomly—it would not serve this function, and the bacterium would starve.

Crucially, this aboutness is simultaneously the origin of value. The *Kriterion* is the exact physical locus where the system registers that one state is better for its persistence than another. The difference between "current state" and "reference state" is not an inert fact; it is a *problem* for the system. If the discrepancy is not closed, the boundary degrades. The *Kriterion* therefore embodies a primitive evaluation: it discriminates between "more conducive to persistence" and "less conducive." Value is not a mysterious addition to life; it is life comparing different states of reality, and acting to close the gap. The comparison *is* the value judgment, at the most fundamental operational level. The protein's shape change is not a metaphor for evaluation; it is evaluation, enacted in the only language that physics understands.

27.3.4 The Origin of the Reference Without a Prior *Kriterion*

The Objection. "You claim the reference state is the condition of the boundary's integrity. But how does a primitive system acquire such a reference without already having a *Kriterion* to determine what 'integrity' means? It seems you need a *Kriterion* to set the reference, but the reference must exist for the *Kriterion* to function. This is a chicken-and-egg problem that unravels your architecture."

The Reply. The derivation does not require that the reference state be cognized or symbolically represented by the system. In the earliest or simplest cases, the reference is a physical attractor—a stable configuration that the system's own dynamics tend to restore. A lipid vesicle that spontaneously reseals after a small tear has an implicit reference: the lowest-energy continuous-membrane state.[3] The vesicle does not "know" what integrity means, but its material properties define a regime of stability, and any departure from that regime triggers passive or active restoration. The reference is not set by a prior *Kriterion*; it emerges from the self-organizing physics of the boundary itself.[4]

As systems become more complex, these implicit attractors are supplemented and eventually replaced by explicit, genetically encoded or learned set-points—ion channels with specific conductances, enzymes with specific binding affinities, neural circuits with specific thresholds. But the logical requirement is not that the reference be established by a prior act of cognition; it is that some physical parameter serves as the

[3]Pier Luigi Luisi, *The Emergence of Life: From Chemical Origins to Synthetic Biology* (Cambridge University Press, 2006).

[4]Deacon, *Incomplete Nature.*

baseline against which current conditions are measured for the purpose of action. The chicken-and-egg dissolves once we recognize that the egg is a physical attractor, and the chicken is the evolutionary refinement of that attractor into an explicit *Kriterion*.[5] No prior mind is required; only the physics of far-from-equilibrium persistence.

27.3.5 The *Kriterion* as the Seat of Value

The Charge. Even if one grants that a *Kriterion* discriminates physical states, one may still object that calling this discrimination "good" and "bad" is an anthropomorphic projection. A voltage difference across a membrane is just a voltage; a chemical gradient is just a gradient. To say the system treats one as good and the other as bad is to import human categories into blind chemistry.

The Reply. This is the crux of the entire derivation, and it must be met without flinching. For an isolated chemical mixture, a gradient is indeed a neutral physical fact. But for a far-from-equilibrium bounded system whose continued existence depends on maintaining that gradient, the gradient is not neutral. If the voltage across the membrane collapses, the cell dies. The *Kriterion*'s registration of the voltage is therefore not merely a physical event; it is a physical event that determines whether the system continues to exist. The stakes are absolute. "Good" is simply the label for the states that, when present, allow the loop to continue; "bad" is the label for the states that, if uncorrected, lead to dissolution. There is no projection here; there is only the causal fact that the system's architecture is organized around avoiding one set of outcomes and seeking another.

The *Kriterion* is the pivot because it is the point where this causal architecture becomes *about* something for the system itself. Before the *Kriterion*, there are only physical variables. At the *Kriterion*, those variables become *evaluated* relative to a reference that is the condition of the system's own existence. This is the minimal, stripped-bare form of value. It does not require consciousness, language, or deliberation. It requires only that the system's internal dynamics are organized such that some states trigger

[5]Daniel C. Dennett, *From Bacteria to Bach and Back: The Evolution of Minds* (W. W. Norton & Company, 2017).

corrective action and others do not. That organization is value, physically enacted. And because the system must act on the basis of its comparisons to persist, those comparisons must correlate well enough with reality. Value, at its origin, is the quality of the fit between the *Kriterion*'s discrimination and the causal structure of the world. This is not poetry; it is physics.

27.4 Against Co-Constitution: The Charge of Uncoupled Architecture

Proposition 6 states that the *Kriterion* and the *Organon*—the model and the action engine—are co-constitutive. They form an inseparable, closed loop where a model cannot exist without action, and directed action cannot exist without a model. The critic objects that this coupling is an artificial pairing, arguing that we can easily find models that never act, and actions that require no model.

27.4.1 The Inert Blueprint

The Objection. "We routinely observe models completely uncoupled from action. A blueprint sitting in an abandoned office contains an incredibly complex, high-fidelity model of a building, yet it performs no work and maintains no boundaries. Similarly, an offline artificial intelligence model stored on a disconnected hard drive contains billions of structural parameters representing human language, yet it sits completely inert. If a model can exist in total isolation from action, then your assertion of co-constitution is empirically false."

The Reply. This objection actually provides the definitive proof for Proposition 6. When viewed through the hard logic of physical persistence, an uncoupled blueprint or a disconnected AI model is not a functional model at all; it is an inert arrangement of matter. The ink on the blueprint is just a spatial distribution of carbon; the AI model is just a static configuration of magnetic or electrical charges.

These artifacts only become "models" when they are functionally coupled to an active agent's comparison-action loop. The blueprint becomes a model when an architect reads it to guide construction; the AI parameters become a model when a processor runs the code to evaluate and generate text. Separated from an active *Organon*, these structures are physically irrelevant to persistence. They are detached structural relics—informational fossils that generate zero directed work (Φ) to offset their own entropic decay (N). A model isolated from action is a dead map with no one to steer the ship; it lacks the circular closure required to constitute an agent's identity.

27.4.2 The Blind Reflex

The Objection. "Look at the lowest levels of biological behavior. A hardwired reflex arc—like a paramecium automatically reversing its cilia upon striking an obstacle, or a human snapping their hand away from a hot stove—executes immediate, directed action without running any internal comparison or possessing a cognitive model. It is pure, mindless, instantaneous reaction. If life can perform boundary-maintaining actions via hardwired reflexes alone, then action does not require a co-constituted model."

The Reply. The critic mistakes processing speed for structural absence. A reflex arc is not the absence of a model; it is a model **pre-computed by evolution and embedded directly into the hardware**.

The structural threshold of the reflex arc is itself a physical comparison operation. In the case of the paramecium, the physical deformation of the front membrane opens mechanosensitive calcium channels. The threshold of those channels *is* the *Kriterion* reference state. If the mechanical pressure exceeds that threshold, the influx of ions forces the cilia to reverse. This is a highly compressed, diachronic model of a physical threat, sculpted over eons of lineage persistence to bypass the metabolic and temporal overhead of real-time deliberation. It is comparison executed at wire-speed, not action without comparison. To say a reflex requires no model is like saying an automated safety valve on a steam pipe operates without a pressure setting. The setting is written into the physical tension of the spring. The reflex is the ultimate validation of co-constitution: the action is so vital that the model has been welded directly into the tool.

27.5 The Wall Holds Again

The *Kybernēsis* core stands. The Primal Distinction is exhaustive; the hurricane, the machine, and the flame have been correctly classified. Directed work is teleonomic, not teleological; spandrels are the tolerated slack around a positive net bias. The cognitive minimum is not semantic inflation; it is a precise functional demarcation grounded in the *Kriterion* loop and the intrinsic reference to boundary integrity. The model and action are co-constitutive in any persistent active engine; apparent counterexamples are either detached fragments (books, blueprints) or fast instantiations (reflexes) of the very architecture in question.

The critic may not be satisfied. The deepest objections lie ahead: to *Hormē* itself, the constitutive imperative, and to the objectivity of value that flows from it. The third wall—Propositions 7 through 11—is where the philosophical stakes become most apparent, and where the critic's metaphysical commitments will be tested most directly. To that wall we turn next.

Chapter 28

The Third Wall: Objections to the Constitutive Imperative, Agency, and Value

The critic has probed the foundations and the middle tiers and found them intact. Now they must confront the conclusions that give the entire edifice its point: the constitutive imperative of *Hormē* (Proposition 7), the continuous *Kybernēsis* imperative (Proposition 8), the isomorphism of life and agency (Proposition 9), the objectivity of value (Proposition 10), and the identity of truth and goodness (Proposition 11). These are the steps that turn a descriptive architecture into a normative one, that fuse physics with ethics, that claim to collapse the Is-Ought gap. They are the most philosophically contentious claims in the book. The critic will bring their strongest weapons here.

This chapter gives the objections to Propositions 7 through 11 their fullest and most adversarial hearing. The charges are grave: conflation of physical necessity with moral obligation, definitional circularity, relativism in value, the failure to account for genuine self-sacrifice, the reduction of truth to utility, and the paradox of the successful evil-doer. Each objection must be met not with dismissal but with genuine engagement, for if any of these charges sticks, the normative project of the derivation unravels.

28.1 Against the Constitutive Imperative: The Physical Is Not the Normative

Proposition 7 states that a far-from-equilibrium system *must* strive to persist, or it ceases to be that system. The "must" is constitutive—it is what it means to be a bounded, non-equilibrium entity over time. The critic sees a conflation.

28.1.1 The Illusion of Option

The Objection. "You claim that *Hormē*—the constitutive drive to persist—is non-negotiable. But I don't experience my existence as a mandatory striving. I can contemplate suicide; I can neglect my health; I can risk my life for a thrill. The bacterium may have no choice but to swim up the gradient, but I clearly do. *Hormē* may describe the bacterium's existence, but not mine. For a human being, persistence feels optional, not constitutive."

The Reply. The feeling of optionality is real, but it is an illusion produced by abundance. A human being is a nested hierarchy of *Hormic* loops, from the ion pumps in every cell to the homeostatic circuits of the brainstem to the learned habits of the *Logistikon* to the abstract self-model of the *Nous*. The lower loops are so robust, so well-buffered, that the highest loop—conscious awareness—rarely feels their urgency. We are fish who have never known water, and the water is *Hormē*.

The proof that the imperative is not optional is available to anyone who attempts to override it. Hold your breath. The urge to breathe begins as a whisper and escalates, within seconds, into a command that no conscious will can resist. You will pass out before you die, and when you pass out, the lower loops—now ungoverned by the *Nous*—will resume breathing on their own. The conscious self can no more opt out of breathing than it can opt out of metabolism. The very brain that contemplates suicide is mortgaged to the oxygen supply maintained by a brainstem that does not ask permission.

The same logic applies to hunger, thirst, pain, and the entire apparatus of *Orexis*. You can refuse food, but the body will consume itself before it dies, and the hormonal scream of ghrelin and the cognitive narrowing of starvation will eventually overwhelm any higher-order intention. The urgency is there, at every level; it is merely gated by the margin of safety that health and abundance provide. A well-fed human in a safe room can entertain the idea that persistence is optional precisely because her lower loops are already securing it so effectively. The bacterium, with one degree of freedom and a razor-thin boundary margin, has no such luxury. It strives or it dissolves, and it knows—in the only way a *Kriterion* can know—that striving is not a choice.

Thus the apparent optionality of *Hormē* at the human level is not evidence against the constitutive imperative; it is evidence of the spectacular success of the lower loops. We are swimming in *Hormē*, and the very fact that we can forget the water is a testament to its depth.

28.1.2 The Is-Ought Conflation

The Objection. "You say the system 'must' strive. But this 'must' is merely a physical necessity: if the system stops striving, it dissolves. This is an 'is,' not an 'ought.' You have not derived a normative imperative; you have described a conditional fact. The statement 'if you stop breathing, you will die' does not imply 'you ought to breathe.' It merely states the consequence. You have smuggled in the normativity by switching from a descriptive 'must' to a prescriptive one. The Is-Ought gap remains unbridged."

The Reply. The derivation does not claim to derive a categorical moral "ought" from a bare physical "is." It claims to derive a *constitutive ought*—an imperative that is built into the very identity of the kind of thing in question. The conditional "if you are a living system, you must strive to persist" is not a moral command; it is an analytic statement about what it means to be a living system over time. It is like saying "if you are a triangle, you must have three angles." There is no gap between "is a triangle" and "has three angles," because having three angles is constitutive of being a triangle.

The Is–Ought gap, as famously articulated by Hume, rested on an ontology of static objects—things like rocks or tables that simply are without having to act to remain so. In such a universe, existence is a passive given; there is no "ought" because there is no "doing." But for a far-from-equilibrium system, existence is a diachronic performance. For these systems, the "is" is not a static state but a continuous achievement maintained by a loop. By failing to distinguish between the passive existence of a rock and the active persistence of an engine, the traditional critique overlooks the fact that for life, the "is" is only sustained by the enactment of the "ought."

The derivation therefore shows that within the category of "far-from-equilibrium bounded systems," there is a constitutive imperative that serves as the anchor for the concept of "good." The derivation does not say "you morally ought to strive"; it says "striving is what you are, and 'good' is the name for the conditions under which your striving succeeds."

The critic may still object: "But why should I care about my own persistence? Even if it is constitutive of my being a living system, I can still ask, 'Is it good that I am a living system?' You haven't answered that." That is Moore's Open Question, and we will meet it shortly. But the constitutive "must" is not the moral "ought" in disguise. It is a precise technical term, and the derivation uses it consistently.

28.1.3 The Suicidal Agent

The Objection. "If *Hormē* is constitutive and non-negotiable, how do you explain suicide? A human being can deliberately end their own life. This is the ultimate rejection of the imperative to persist. If a system can act against its own *Hormē*, then *Hormē* is not constitutive—it is merely a strong drive that can be overridden, like hunger or fear. Your universal imperative has a glaring counterexample."

The Reply. Suicide is not a refutation of *Hormē*; it is a demonstration of *Hormē* operating at a higher-order boundary, or of a catastrophic failure of the model.

In the first case, the individual who sacrifices their life for a cause—a soldier falling on a grenade, a martyr accepting death—is not abandoning *Hormē* but redirecting it. Their highest-order boundary, the identity with which they are ultimately concerned, is not the individual body but the collective—the family, the nation, the faith. The sacrifice preserves the higher-order boundary at the cost of the lower. *Hormē* is scale-invariant; it operates wherever a boundary is maintained. The martyr maintains a boundary of value, of meaning, of identity; the physical death is the ultimate redirection of directed work from the somatic to the symbolic. This is not an exception to *Hormē* but a profound expression of it.

In the second case, the suicide results from a catastrophic failure of the *Kriterion*'s model. The individual who takes their life because of depression, despair, or psychosis is acting on a map that is profoundly decoupled from causal reality. They believe, falsely, that their future holds only suffering, that they are a burden, that their death will improve things. These are epistemic errors of the most tragic kind. The framework predicts that when a *Kriterion*'s model is fatally inaccurate, the boundary will dissolve—and it does. The suicide is not evidence that *Hormē* is optional; it is evidence that a broken *Kriterion* leads to dissolution, exactly as the truth-goodness isomorphism (Proposition 11) would predict.

The critic might press: "But what of the rational suicide—the person who, facing terminal illness, chooses assisted death with full clarity and accurate expectations?" This too is nested *Hormē*. The person is maintaining the boundary of dignity, of self-authorship, of a life that meets their constitutive standards of what their existence should be. When the body can no longer sustain those standards, the higher-order boundary—the self-concept—chooses its own terminus. The derivation does not condemn this; it explains it.

28.1.4 The Machine's Missing Drive

The Objection. "A solar-powered pump senses, compares, and acts to maintain a water level. It is an active engine by your earlier criteria. Does it possess *Hormē*? If it does, then a water pump has a constitutive drive, which is absurd. If it does not, then what is the difference, and how is that difference derived rather than stipulated?"

The Reply. The pump lacks intrinsic *Hormē* because its *Kriterion*'s reference state—the desired water level—is not its own boundary condition. The pump does not maintain the pump. If the pump's solar panel cracks, it does not repair itself. If its motor burns out, it does not order a replacement. Its loop is organized around a goal that is external to its own persistence as a bounded entity. Its *Hormē*, if the term can even be applied, is *extrinsic*—it is a fragment of the human designer's *Hormē*, extended into machinery.

The bacterium's chemotaxis loop, by contrast, is organized around the maintenance of the bacterium's own boundary. The reference state—the ion gradients, the membrane integrity—is the condition of the bacterium's own continued existence. The work the bacterium performs is self-directed. This is intrinsic *Hormē*.

The distinction between intrinsic and extrinsic *Hormē* is not an arbitrary add-on. It follows from the derivation: a bounded system whose *Kriterion* reference is the condition of its own boundary is a system whose loop is self-referential. Its striving is for itself. A bounded system whose *Kriterion* reference is externally imposed and whose boundary is maintained by an external agent is a tool, a derived agent. The pump is the latter; the bacterium is the former. The distinction is crisp, grounded, and explanatory.

28.2 Against the Life-Agency Isomorphism: The Charge of Circularity and Counterexamples

Proposition 9 asserts that life and minimal agency are isomorphic—a system is alive if and only if it possesses *Hormē*, and it possesses *Hormē* if and only if it is an agent. The critic advances several counterexamples designed to pry life and agency apart.

28.2.1 The Virus

The Objection. "A virus is a standard problem case. It evolves; it has a boundary (a capsid); it contains information; it redirects the host cell's machinery to produce more viruses. By many criteria, it looks like an agent. Yet you classify it as non-living, a passive particle. Why? If the virus redirects the causal flow toward its own replication, is that not minimal agency? And if it has minimal agency, why is it not alive? Your isomorphism seems to break down exactly at the most contested border."

The Reply. The virus is not an agent because it has no intrinsic *Hormē*. In its extracellular phase, the virus is a passive particle. It performs no directed work; it has no *Kriterion* loop; it does not maintain its own boundary against disruption. It drifts. Its capsid is not actively maintained; it is a static protein shell that protects the genetic cargo until it happens to encounter a host cell. The virus does not swim, does not metabolize, does not repair.

Upon entering a host cell, the virus hijacks the host's *Hormē*. The viral genetic material inserts itself into the host's *Kriterion-Organon* machinery, reprogramming the host's metabolic pathways to produce viral particles. The host's directed work now flows toward viral replication. The agency, such as it is, belongs to the host-parasite complex, not to the virus alone. The virus is an informational parasite—a fragment of genetic instruction that exploits agency without being an agent. It is not alive for the same reason a computer virus is not alive: it is a piece of code that runs on hardware maintained by something else.

The critic might ask: "But what about the virocell—the infected cell, which is a new kind of entity, a host-virus hybrid? Is that alive?" Yes, the infected cell is alive—it is an active engine with a corrupted *Kriterion*. Its *Hormē* has been redirected. But the virus particle outside the cell is not. The framework's ability to make this distinction—virus particle as passive, virocell as living—is a strength, not a weakness.

28.2.2 The Chess AI and Derived Agency

The Objection. "A chess AI evaluates positions, compares them to a winning criterion, and selects moves. It satisfies the operational criteria: it discriminates, compares, and acts. It is an agent. It is not alive. Therefore, agency and life are not isomorphic."

The Reply. The chess AI possesses *derived agency*, not intrinsic agency. Its winning criterion is externally programmed; its physical persistence depends on external power and cooling; its boundary is not a self-maintaining far-from-equilibrium system. The AI is a tool—an extension of the human designer's *Hormē*. It is an agent only in the sense that a guided missile is an agent: it steers toward a goal, but the goal is not its own.

The life-agency isomorphism is a claim about *intrinsic* agency. It states that any system that is *intrinsically* an agent—any system whose *Kriterion* reference is its own boundary integrity—is alive, and vice versa. The chess AI is not an intrinsic agent; it is a derived agent. The isomorphism does not apply to derived agents. The distinction is clear: intrinsic agency requires self-maintenance of the boundary that houses the loop. Derived agency does not.

28.2.3 Definitional Circularity

The Objection. "You define life in terms of *Hormē* and agency in terms of *Hormē*. This is a definitional circle. You have not discovered an isomorphism; you have stipulated one by assigning both terms to the same underlying property."

The Reply. The definitions were built independently. Life was defined as the internal architecture of maintenance (the *Kriterion-Organon* loop), while Agency was defined as the external redirection of causal flow. The convergence of these two descriptions onto a single architecture is a theorem of the derivation, not a stipulation.

It would have been logically possible for the two to diverge. We can imagine two "failed" states where the isomorphism breaks:

1. **Life without Agency (The Static Trap):** If the universe permitted a system to persist indefinitely through purely passive material stability—like an eternal, unbreakable crystal—we would have a form of "Life" that never acts. Such a system would have a boundary but no need for *Hormē.* But the Second Law of Thermodynamics forbids this; in a disruptive world, no material is truly "impervious." Persistence requires active repair.
2. **Agency without Life (The Dissipative Trap):** We can observe systems that redirect causal flow with staggering power—like hurricanes or wildfires—but which possess no boundary to maintain for their own sake. These are "active" in a sense, but they are not agents because their actions are directed toward the acceleration of their own dissolution, not their persistence.

Propositions 2 and 4 close these paths. They establish that in a far-from-equilibrium universe, **maintenance is action** and **action requires a boundary.** The discovery is that the exact same mechanical loop required to solve the problem of *staying alive* (Life) is the exact same loop required to *change the world* (Agency). The isomorphism is not a definitional fiat; it is an ontological entailment. The definitions of Life and Agency remain conceptually distinct, but in a disruptive universe, they are physically inseparable.

28.2.4 The Anthropocentric Charge

The Objection. "Calling a bacterium an 'agent' cheapens the term. Agency, in any meaningful sense, involves deliberation, conscious choice, and the capacity to do otherwise. A bacterium is a biochemical automaton, blindly executing chemical routines. Your 'minimal agency' is a rhetorical device that projects human dignity onto mindless chemistry."

The Reply. The objection assumes that agency must be defined by its most elaborated human form and then denies the term to anything simpler. The derivation rejects this on methodological grounds. If agency is a natural phenomenon rather than a supernatural gift, it must have emerged from simpler precursors, and those precursors must share the same fundamental causal structure—the capacity to redirect causal flow based on discrimination of states.

A bacterium is not a biochemical automaton in the sense of a passive mechanism whose output is mechanically determined by its input. It discriminates among possible states and alters its behavior accordingly. If the chemical gradient reverses, it reverses its swimming bias. It *can* do otherwise, in the only sense physics requires: its internal dynamics are such that different environmental conditions produce different outputs, and these outputs alter the probability distribution of its future states in ways that favor persistence. That is the operational core of agency, present in the bacterium as surely as the seed is present in the oak.

Furthermore, if we define agency as a transcendental property—one that requires a "soul," a "will," or a "consciousness" that is fundamentally decoupled from physical mechanism—we have moved beyond the realm of falsifiable theory. Such a definition is a conversation-stopper; it asserts a metaphysical gap that no amount of causal evidence could ever bridge. The derivation, by contrast, provides a testable, physical account of how the minimal form scales into the reflective form.[1] The same *Kriterion* architecture that steers a bacterium, when recursively applied to the system's own models, yields the capacity for self-governance and deliberate choice. The highest expression of agency presupposes the minimal form, not the other way around. The bacterium's agency is not a metaphor; it is the root of the tree whose highest branch is human deliberation.

28.2.5 The Metabolic Mutant

The Objection. "Consider a hypothetical cell that has lost all sensory apparatus—no receptors, no signaling cascades—but still performs metabolism, grows, and divides. It would be alive by any intuitive standard. But it would have no *Kriterion*, and therefore no agency—and, by your criteria, no *Hormē*. It would refute the isomorphism."

[1] For a complementary empirical framework showing how agency scales continuously from cellular homeostasis to macro-level cognition, see Michael Levin, "The Computational Boundary of a 'Self': Developmental Bioelectricity as a Bottom-up Mechanism for Biological Systems to Create Granular Scale-Free Intelligence," *Frontiers in Psychology* 10 (2019): 2688.

The Reply. The critic is imagining something that cannot exist. A cell cannot perform metabolism without a selectively permeable membrane. That membrane is itself a sensor: its passive permeability to certain ions and its active transport of others constitute a physical comparison between the cell's internal state and the environment. Even the simplest cell maintains ion gradients because the membrane is selectively permeable, and the leakage through that membrane is compared, in effect, to the pumping rate that counteracts it. There is no metabolism without a boundary, and there is no boundary without a differential interaction with the environment. That differential interaction is the minimal *Kriterion*.

The "metabolic mutant with no sensory apparatus" is a philosopher's fiction, like a triangle with four sides. If it truly lacked all sensitivity to its environment, it would be a sealed inert compartment, and it would equilibrate with its surroundings and die. The critic must produce a real example of a living system with zero discrimination. None exists.

28.3 Against the Objectivity of Value: Relativism, Nihilism, and the Naturalistic Fallacy

Proposition 10 asserts that value is objective: "good" is that which fulfills a system's *Hormē*, and "bad" is that which frustrates it. This is the claim that most directly challenges the philosophical orthodoxy of the last century. The critic brings the heavy artillery.

28.3.1 Relativism

The Objection. "Under your framework, what is good for the spider is eating the fly; what is good for the fly is escaping. The same event—the spider's web catching the fly—is simultaneously good and bad, depending on the system. This is not objective value; it is relativism dressed in physicalist language. You have not discovered an objective good; you have discovered that goods are system-relative, which is precisely the problem with grounding ethics."

The Reply. The objectivity of value does not require that there be a single state of affairs that is good for all possible systems at all possible times. It requires only that, for any given system, the goodness of an outcome is determined by objective, physical facts about that system's organization and its relationship to the environment. Relativity to a system does not imply subjectivity, any more than relativity of motion to a reference frame implies that motion is "subjective."

The derivation is entirely compatible with genuine conflicts of value. The spider and the fly have irreconcilable goods; the predator and the prey are locked in a zero-sum game. This is not a failure of objectivity; it is a recognition that the universe contains genuine normative conflict—a fact that any adequate meta-ethics must accommodate. The derivation explains why such conflicts arise (different systems have different persistence conditions) and provides a diagnostic for analyzing them (trace the nested architecture of the systems involved and examine the objective causal dependencies). It does not promise harmony; it promises clarity.

The critic's demand that an objective good must be a *universal* good is a hangover from a monotheistic or universalist metaphysics that the derivation explicitly rejects. Value is plural because entities are plural. The good of the bacterium is not the good of the human, and neither is the good of the stone, because the stone has no good. This is not relativism; it is pluralism grounded in physical identity.

28.3.2 The Nihilist's Performative Contradiction

The Objection. "A person can look at your entire derivation and shrug. 'I don't care about my persistence,' they say. 'I assign no value to it. Nothing matters to me.' Your derivation cannot refute the nihilist, because the nihilist simply opts out of the whole game. Value is not objective if it can be coherently rejected."

The Reply. The nihilist's rejection is verbally coherent but physically incoherent. The nihilist who says "nothing matters, not even my own persistence" is saying it with a mouth that is metabolizing glucose, with a brain that is regulating its own temperature, with a body that is continuously performing directed work to maintain its boundary. The nihilist's body, at every level below the utterance, is in the business of treating certain states as worth achieving and others as worth avoiding. The utterance is a performative contradiction: an act produced by a system that is, in its operation, committed to the objectivity of its own persistence conditions.

The nihilist can, of course, choose to act against their own persistence—they can stop eating, stop drinking, walk into the sea. But that action does not demonstrate that value is subjective; it demonstrates that the nihilist's higher-order *Kriterion* has adopted a terminal reference state. They are still acting on a value—the value of demonstrating their nihilism, the value of ending suffering, the value of asserting their will—even as they deny that value exists. The very act of "opting out" is an expression of *Hormē* at some level of their nested identity.

The derivation does not force the nihilist to care. It simply notes that the nihilist already cares, operationally, at every level below the philosophical one. The verbal denial of value is a luxury afforded by the very metabolic striving it denies.

28.3.3 The Naturalistic Fallacy

The Objection. "Hume's law states that you cannot derive an 'ought' from an 'is.' You have defined 'good' as 'that which fulfills *Hormē*.' But *Hormē* is an 'is'—a physical fact about a system's organization. You have simply stipulated that this 'is' counts as a good. That stipulation is not a derivation; it is a redefinition. The Naturalistic Fallacy is not dodged by redefining the target term."

The Reply. The derivation does not commit the Naturalistic Fallacy because it does not attempt to derive a *categorical* moral ought from a bare physical is. It identifies a *constitutive* relationship within a specific class of systems—far-from-equilibrium, bounded entities—and notes that this relationship is the only possible non-mysterious anchor for the concept of "good" as applied to those systems. The claim is meta-ethical, not directly ethical. It says: if you want to understand what "good" means for a living thing, the answer is *Hormē*-fulfillment, because that is the operational standard that the thing itself instantiates.

The critic who insists that "good" must mean something else—something that issues categorical commands to all rational beings regardless of their constitution—is welcome to provide that account. The derivation's claim is that no such account has ever succeeded in grounding its commands in the actual, causal structure of the world. Theistic ethics grounds them in divine will; Kantian ethics grounds them in the structure of practical reason; utilitarian ethics grounds them in the felt quality of pleasure. Each of these leaves a gap between the grounding and the grounded. The derivation closes the gap by identifying the good with the very condition of there being an agent who can ask the question. That is not a fallacy; it is a solution.

28.4 Against the Identity of Truth and Goodness: The Successful Vice

Proposition 11 asserts that epistemic error and ethical vice are functionally identical—both are misalignments of the model with the causal structure of reality. The critic brings forth the specter of the successful psychopath, the martyr who dies for truth, and the fitness-beats-truth paradox from evolutionary game theory. If truth and goodness are the same, how can vice succeed? How can truth lead to death? How can falsehood aid survival?

28.4.1 The Successful Psychopath

The Objection. "Consider a ruthless dictator who exploits, tortures, and kills, and yet dies peacefully in old age, surrounded by wealth and power. Their model of the world—that trust is for fools, that empathy is a weakness—was, by all evidence, accurate *for them.* It kept them alive, powerful, and comfortable. According to your derivation, their actions were 'good' because they fulfilled their individual *Hormē.* This is not a reductio; it's a moral abomination. If your derivation cannot condemn the successful psychopath, it is ethically bankrupt."

The Reply. The derivation distinguishes between the basal mechanics of value and the higher-order ethical obligations that arise from nested social boundaries. The derivation establishes that the good for any system is *Hormē*-fulfillment; it does not, by itself, generate a complete human ethics. That requires additional considerations about the structure of social persistence.

The successful psychopath's persistence is possible only within a social environment where most people are *not* psychopathic. Trust, cooperation, and empathy are the norm; the psychopath parasitizes this norm. Their strategy is frequency-dependent: if it were universal, the social boundary—the collective *Hormē* of the society—would dissolve, and all would perish. The psychopath's success is therefore parasitic on the very values they reject.

Moreover, the derivation provides the tools to condemn the psychopath from within any adequately nested perspective. From the perspective of the society, the psychopath is a threat to the collective boundary and thus objectively bad for the society. From the perspective of the psychopath's own long-term flourishing—if they cared about legacy, about genuine human connection, about anything beyond raw power—their strategy is catastrophic, because it forecloses the higher-order goods that make human life meaningful. The derivation can say all of this without appealing to a mysterious moral property. It can say: the psychopath's model is causally inaccurate with respect to the nested boundaries they ignore, and this inaccuracy leads to a life that is, by any measure of full human flourishing, a failure.

The critic may still feel a residue of dissatisfaction: "But what if the psychopath doesn't care about any of that? What if their only good is power, and they achieve it?" Then the derivation says: for that individual, given their actual (impoverished) reference states, their actions fulfilled their *Hormē*. The derivation does not pretend that every agent's *Hormē* is equally rich, equally admirable, or equally conducive to a flourishing society. It observes that some agents have broken *Kriteria*—impoverished models of the good—and that their success, while real on their own terms, is parasitic and ultimately fragile. That is not a failure of the derivation; it is a recognition of the tragic structure of value in a world of nested, conflicting agents.

28.4.2 Martyrdom: Dying for Truth

The Objection. "A martyr dies for a true cause—say, a scientist who speaks truth to a tyrannical power and is executed. The scientist's model was accurate; the cause was just. Yet the outcome was death—the ultimate failure of individual *Hormē*. Truth did not lead to goodness; it led to destruction. Your isomorphism is broken."

The Reply. The martyr's death is a nested *Hormē* triumph. The scientist's highest-order boundary was not their individual body but the truth itself, the scientific community, the future of knowledge. By refusing to recant, they preserved that higher-order boundary—the integrity of their identity as a truth-seeker—even as the lower-order boundary dissolved. If the cause was just and the sacrifice contributed to its eventual victory, then the martyr's model was accurate, and their action fulfilled *Hormē* at the level that mattered most to their self-model.

If the sacrifice was futile—if the tyranny continued unchallenged, if the truth was suppressed and forgotten—then the martyr's model was in fact inaccurate. They acted on a false prediction that their death would make a difference. The derivation says that when a model is inaccurate, the boundary dissolves, and the good is not achieved. A futile martyrdom is a tragedy of epistemic failure, not a refutation of the isomorphism. Truth and goodness remain aligned: accurate nested models yield genuine preservation; inaccurate ones yield dissolution.

28.4.3 Addiction and Akrasia: The Decoupled Will

The Objection. "An addict can possess a perfectly accurate belief that the drug is destroying their life, and yet continue to use it. Their model is true; their action is bad. This seems to decouple truth and goodness entirely. The isomorphism is false."

The Reply. The addict is not a monolithic agent. The *Psyche* is a stratified architecture of nested *Hormic* loops, as described in Chapter 20: Nested Architecture and Conflict Resolution. The lower-order loop—the neurochemical reward pathway—has been recalibrated by the drug to treat the substance as the highest good. Its *Kriterion* is accurate *relative to its own corrupted reference state*: it correctly identifies the substance that will satisfy the recalibrated receptor. The higher-order loop—the cognitive model—knows that this satisfaction is destroying the macro-boundary.

The conflict is not between truth and falsehood at the same level; it is a civil war between two models, one of which is catastrophically false about the overall persistence conditions of the whole. The truth-goodness isomorphism applies to the integrated system: when the higher-order model succeeds in governing the lower-order loop, truth and goodness converge in recovery. When the lower-order loop dominates, the system's overall model is operationally false (it acts as if the drug is vital), and the boundary degrades. The existence of the conflict is not a counterexample; it is an illustration of the nested architecture of *Hormē*, and the catastrophe that occurs when a local model usurps the global one.

28.4.4 The Fitness-Beats-Truth Paradox

The Objection. "Evolutionary game theory has shown that organisms can evolve to track 'fitness-relevant interfaces' rather than objective reality. A prey animal that hallucinates a predator at every shadow may have a higher fitness than one that calculates the true probability. Fitness selects for useful illusions. Your claim that survival requires truth is empirically false."[2]

The Reply. This objection was addressed in Chapter 18: The Epistemology of Asymmetric Risk (The Epistemology of Asymmetric Risk). The prey animal's "hallucination" is not a falsehood about the causally relevant variable. The causally relevant variable is not the visual identity of the stimulus (shadow vs. predator); it is the *expected cost of the decision under uncertainty.* The prey animal's model is a compressed, operationally accurate map of the causal stakes: a false negative costs everything, a false positive costs a trivial metabolic outlay. The model that triggers flight on ambiguous stimuli is *true* with respect to the causal risk structure, even if it is not representationally faithful to the visual field.

The derivation's concept of truth is causal and operational, not representational in the narrow sense of a photograph. A map is true if it guides actions that maximize the expected value of the boundary integral. The "paranoid" prey animal's map is true in this sense; the "rational" prey animal's map, which ignores the asymmetry of stakes, is false. The fitness-beats-truth paradox dissolves when truth is properly understood as the accurate tracking of the causal structure of reality, weighted by the stakes that reality imposes on the agent's boundary.

28.5 The Wall Still Stands

The third wall has been struck by the heaviest philosophical artillery available. The Is-Ought conflation was clarified as a constitutive, not a moral, imperative. The suicidal agent was shown to be an expression of nested *Hormē*, not a refutation. The virus and the chess AI were correctly classified as non-living, non-intrinsic agents. Moore's Open Question was answered by a transcendental argument that reveals the asker's own constitutive commitment. Relativism was shown to be a feature, not a bug—a

[2]Donald D. Hoffman et al., "The Interface Theory of Perception," *Psychonomic Bulletin & Review* 22, no. 6 (2015): 1480–506.

pluralism of objective goods grounded in distinct physical identities. The nihilist was caught in performative contradiction. The naturalistic fallacy was dodged by showing that the derivation does not export a bare "is" to a categorical "ought," but identifies the constitutive conditions under which "good" has a non-mysterious meaning.

The successful psychopath was contextualized within nested social boundaries. The martyr's death was reclassified as a triumph at a higher order, or a tragedy of epistemic failure. Addiction was reframed as a civil war between nested models. The fitness-beats-truth paradox was dissolved by clarifying truth as causal-operational accuracy.

The localized, systemic defenses of the paradigm are complete. But the critic has one final line of assault. Having failed to break the normative core at the level of individual biological and behavioral systems, the skepticism now scales upward to challenge the structural integrity of the entire paradigm. The next chapter will confront framework-level objections—interrogating our foundational redefinition of teleonomy, the substrate-neutral deduction of natural selection, the mathematical validity of the boundary integral, and the cosmological mandate of a deterministic universe. The philosophical core has been defended. The wall, battered but unbreached, remains.

Chapter 29

The Fourth Wall: Framework-Level Objections

The third wall defended the normative heart of the derivation—the leap from physical persistence to objective value. The critic tested the collapse of the Is-Ought gap and the objectivity of the good, and the logic held. But a framework of this scale makes claims that stretch across multiple disciplines. It redefines purpose without teleology, claims Darwinian evolution as a deductive theorem, formalizes survival into a mathematical integral, reinterprets cognitive bias as epistemological truth, demands an ordered universe, and constructs a nested hierarchy of the mind.

The critic, having failed to break the normative core, must now test these structural pillars. This chapter constitutes the fourth wall: the framework-level objections. Here, we defend the explanatory reach of the derivation against the sciences and philosophies it intersects, proving that its deductive consequences are as robust as its foundational axioms.

29.1 Objections to Teleonomy

The framework draws a sharp distinction between teleology (future goals causing present action) and teleonomy (present architecture shaped by past persistence). This distinction is the backbone of the claim that directed work is genuinely directive without being purposive in any mystical sense. Four objections challenge the coherence of this distinction.

29.1.1 The Teleology Charge

The Objection. "The entire framework is teleological in disguise. *Hormē* is called an 'imperative'; the system 'must' strive; work is 'directed' toward persistence. These are the idioms of purpose. The distinction between teleology and teleonomy is a semantic trick: you have simply relabeled what everyone else calls teleology and declared the problem solved."

The Reply. The "must" of *Hormē* is constitutive, not prescriptive. It is the same "must" as "a bottomless bucket must be continuously filled or it becomes empty." No future state pulls the bucket; the bucket's structure and the continuous drain impose the necessity. Similarly, the system's architecture—its far-from-equilibrium boundary and blind environment—imposes the requirement of continuous directed work. The language of striving is functional shorthand for the operation of the *Kriterion-Organon* loop, fully translatable into proximate causal terms. The difference between teleology and teleonomy is not a choice of words; it is a choice of causal model. In teleology, a future goal explains present action. In teleonomy, present structure explains present action, and past selection explains the structure. Those are two entirely different explanatory architectures, and the framework consistently occupies the second.

29.1.2 The Designer Objection (The Watchmaker Returns)

The Objection. "If teleonomy requires a reference state that guides action, and the reference state is exquisitely tuned to the environment, then the existence of such a state seems to demand an intelligent designer. The classic watchmaker argument—that a watch implies a watchmaker—applies with equal force to the *Kriterion* and its set-point. You cannot explain away the appearance of design by invoking a blind process unless you can show that the process is genuinely capable of producing the design."

The Reply. The watchmaker argument trades on an analogy that the derivation systematically dismantles. A watch is a tool; its reference state (the time it must display) is externally imposed by a designer. The watch does not maintain its own boundary; it does not repair its own gears; it does not strive. An organism's reference states are constitutive, not imposed—they are the physical conditions of its own boundary integrity. Moreover, the derivation does not merely *assert* that blind processes can produce functional complexity; it *deduces* that they must. Chapter 16: Darwinian Evolution as a Diachronic Consequence demonstrates that natural selection—the differential persistence of variants whose *Kriterion-Organon* loops more effectively bias outcomes toward boundary maintenance—follows deductively from finite lifespans, replication with variation, and the constitutive imperative. No designer is required because the logic of differential persistence operating over deep time is, in principle, sufficient to generate the complexity we observe. The watchmaker argument depends on the claim that natural selection *cannot* do this; the derivation shows that it can, and does.

29.1.3 The Semantic Objection

The Objection. "You have simply replaced 'teleology' with 'teleonomy' and continued to use the same goal-directed language—directed, strive, good, bad. This is a cosmetic change. If the language is identical, the underlying concept is identical. You are practicing intellectual dishonesty by pretending to have solved a problem you have merely renamed."

The Reply. The identity of the language is superficial; the underlying causal model is different in kind. When a physicist says a ball "seeks" the lowest point in a gravitational well, no one accuses physics of teleology, because the full causal story—gradient forces producing acceleration—is available. The framework's functional language is exactly analogous. "The bacterium strives toward glucose" is shorthand for: chemoreceptors bind glucose; the *Kriterion* detects a deviation from baseline; the *Organon* biases the motor; the cell moves up the gradient. Every step is a physical event, fully specified by efficient causes. The shorthand is legitimate because it picks out the pattern in the causal chain that is relevant to understanding persistence—just as "seeks" picks out the pattern of motion toward a local minimum in a potential field. To insist that the framework is teleological because it uses functional language is to indict every science from physics to biology for the same crime.

29.1.4 The Origin of Teleonomy (The Circularity Charge)

The Objection. "You claim that a teleonomic system requires a *Kriterion* with a reference state, and that the reference state is the product of past selection. But the first teleonomic system could not have benefited from past selection—there was no past. How did the first *Kriterion* arise? If you cannot answer, your account of teleonomy presupposes the very thing it seeks to explain."

The Reply. This is the objection from abiogenesis, and it does not threaten the logical structure of teleonomy; it identifies an empirical question about the historical origin of the first such system. The derivation does not claim to describe the specific chemical pathway from prebiotic matter to the first *Kriterion.* It claims only that any persisting active engine must possess the architecture described, and that this architecture is logically compatible with a self-organizing origin.

A lipid vesicle with a preferred membrane curvature exhibits a primitive "reference state" without any prior selection—the curvature is a physical attractor of the system's dynamics. If fluctuations distort the membrane, the system relaxes back to the attractor, performing something functionally equivalent to striving without any evolutionary history. Once such a system replicates with variation, natural selection can refine the attractor into an explicit set-point, and the full teleonomic architecture emerges.[1] The circle is not vicious; it is a historical sequence: physical attractor → primitive teleonomy → replication → selection → explicit *Kriterion.* The derivation requires only that this sequence is possible, and chemistry provides abundant grounds for thinking that it is.[2]

29.2 Objections to the Diachronic Deduction of Natural Selection

The extension of *Hormē* across generations (Chapter 16: Darwinian Evolution as a Diachronic Consequence) argues that natural selection follows deductively from finite lifespans, replication with variation, and the constitutive imperative. Four objections challenge the scope and independence of this claim.

29.2.1 Replication Is Not Deduced

The Objection. "Your deduction of natural selection depends on the empirical fact that organisms replicate. But replication is not derived within Propositions 1–11; it is an additional premise imported from observation. Therefore, natural selection is not a purely deductive consequence of the definition of life; it depends on a contingent fact about the world."

The Reply. The framework accepts this point without embarrassment. The deduction is conditional: *if* finite agents replicate with variation, *then* natural selection follows with mathematical inevitability. Replication is indeed an empirical observation, as is the finitude of individual lifespans.

[1]Ernst Mayr, "Teleological and Teleonomic: A New Analysis," *Boston Studies in the Philosophy of Science* 14 (1974): 91–117.

[2]Jacques Monod, *Chance and Necessity: An Essay on the Natural Philosophy of Modern Biology* (Alfred A. Knopf, 1971).

The derivation does not pretend to conjure the existence of reproduction from pure logic; it shows that, *given* the observed facts of death and replication, the machinery of differential persistence is not an additional hypothesis but the diachronic shadow of *Hormē* itself. Moreover, replication is not a random add-on—it is the only available strategy for an organizational pattern to persist beyond the dissolution of its current physical vehicle. The logic explains why replication exists, even if it does not deduce its existence *a priori*.

29.2.2 The Origin of Life Gap

The Objection. "Your deduction explains how existing life evolves, but says nothing about how life originated. If the framework cannot explain abiogenesis, it is incomplete."

The Reply. A theory of the necessary architecture of life is not obligated to provide a historical narrative of its first instantiation. The framework defines the target: any viable origin-of-life scenario must produce a bounded, far-from-equilibrium system possessing a minimal comparison-action loop—the architecture of Propositions 1–8. How nature first achieved this is an empirical question, and the framework is compatible with multiple pathways (hydrothermal vents, tidal pools, lipid-world scenarios). The derivation's completeness lies in its logical structure, not in its possession of a time-machine. Setting the target is, itself, a contribution: it tells empirical science what to look for.

29.2.3 Lamarckian Inheritance Does Not Undermine the Deduction

The Objection. "If acquired traits can be inherited—through epigenetic modification, cultural transmission, or other non-genetic channels—then evolution deviates from the standard neo-Darwinian model. This undermines the claim that natural selection is a deductive consequence of the framework, because the framework appears tied to a specific (and perhaps outdated) mechanism of variation."

The Reply. The term "natural selection" in the derivation refers exclusively to the core principle of differential persistence of heritable variants, not to any specific mechanism of variation. Darwin formulated the principle without knowledge of genes; his logic was substrate-neutral. The deduction is equally substrate-neutral. Genetic mutation, epigenetic modification, horizontal gene transfer, cultural innovation—all are channels of variation, and the logic of differential persistence applies to them all.[3] The framework predicts that, wherever multiple channels compete, the one—or the combination—most effective at preserving *Hormic* patterns will be differentially retained. The recent surge of interest in epigenetics and niche construction does not threaten the deduction; it confirms the expectation that inheritance mechanisms will be richer than any single chemical commitment could anticipate.

29.2.4 Neutral Evolution and Drift

The Objection. "Much of evolutionary change is neutral—genetic drift, spandrels, and random fixation of variants that have no effect on fitness. If evolution is the diachronic expression of *Hormē*, why is there so much tolerated noise? A directed force should produce a tighter signal."

The Reply. The derivation claims that natural selection is *a* consequence of replication with variation in a disruptive environment, not the *only* consequence. Drift is possible precisely because a system can tolerate neutral variation as long as the net bias toward persistence remains positive. Proposition 4 states that not every action need be optimal; the system must only remain above the dissolution threshold. Diachronically, a lineage can accumulate neutral and even mildly deleterious traits as long as its overall *Hormic* capacity stays adequate. Drift is the tolerated noise around the directed signal of selection—evidence not of the absence of direction, but of a margin of safety that permits variation without immediate collapse.

29.2.5 The Parochial Strategy Objection (Is Replication Truly Necessary?)

The Objection. "You argue that natural selection is a deductive consequence because a singular boundary will inevitably encounter an environmental disruption (N) exceeding its work capacity (Φ) over deep time. Therefore, you claim replication is mathematically necessary to distribute boundary risk. But this assumes replication is the *only* physical solution to deep-time disruption. What about an organism that achieves perfect

[3] Eva Jablonka and Marion J. Lamb, *Evolution in Four Dimensions: Genetic, Epigenetic, Behavioral, and Symbolic Variation in the History of Life*, Revised (MIT Press, 2014).

regenerative durability and modular repair, allowing it to survive forever without breeding? By elevating replication to a structural necessity, you are conflating a specific, parochial biological strategy with a universal physical law. The anthropic proof only shows that *Earth* life replicated, not that replication is deductively required for *any* active engine."

The Reply. The framework's deduction of replication is topologically and thermodynamically absolute; it is not a biological accident. The critic's alternative strategy—perfect modular regeneration—fails on two strict mathematical fronts: one of physical capacity, and one of statistical origin.

First, consider the boundary integral. For any finite system, its maximum directed work capacity (Φ) at any given moment t is strictly finite. Over geological deep time, the probability of a localized, catastrophic N-spike (e.g., a meteor strike, a nearby supernova, total thermal collapse) exceeding that finite Φ approaches absolute certainty. Perfect modular regeneration does not solve this; a catastrophic N-spike will simultaneously vaporize the regenerative machinery and the boundary it maintains. Because the probability of a singular, contiguous boundary failing over deep time is 100%, the only physical defense against localized annihilation is spatial redundancy. The active engine must create parallel, independent boundaries that do not share the exact same spatial vulnerability. This distribution of independent boundary risk is the precise definition of reproduction.

Second, consider the origin of this hypothetical perfect durability. A single organism possessing the flawlessly adapted, complex "modular fit" required to neutralize all terrestrial N-spikes cannot emerge *ex nihilo*. The spontaneous assembly of such a highly ordered, statistically improbable system by pure chance is effectively zero—an event mathematically indistinguishable from a theological miracle. The only physical, non-miraculous mechanism capable of climbing the probability gradient to achieve complex adaptation is the iterative, differential retention of functional variants. That iterative algorithm *is* reproduction and natural selection.

Therefore, the "perfectly durable" single organism is a logical impossibility. It cannot survive a localized catastrophic N-spike, and the very complexity required to attempt such durability can only be built by a reproducing lineage. Reproduction is the sole topological solution to the calculus of deep time.

29.3 Objections to the Boundary Integral

The boundary integral (Chapter 17: The Mathematics of Boundary Maintenance) formalizes the constitutive condition for persistence. Three objections challenge its novelty, measurability, and applicability to systems without sharp boundaries.

29.3.1 The Equation Is a Thermodynamic Truism

The Objection. "Your equation $A(t) = A_0 + \int(\Phi - N)d\tau$ says nothing beyond the Second Law of Thermodynamics: order decays unless energy is expended to maintain it. This is not new; it is a restatement of entropy dressed in mathematical notation."

The Reply. The Second Law is a special case for thermodynamic entropy in closed systems. The boundary integral is a general constitutive condition for any organized pattern, regardless of substrate. It applies even when the disruptive force is not thermodynamic—mechanical erosion, chemical attack, information degradation. Moreover, the integral specifies the *functional* character of the countervailing work: Φ must be *directed* by a *Kriterion* loop that tracks the boundary's integrity. A heat engine dumping waste heat does not maintain its own boundary; the integral requires a specific coupling between Φ and $A(t)$—the Kybernēsis—which is not derivable from thermodynamics alone. The integral is a logical formalization of the constitutive imperative, not a physical law.

29.3.2 The Variables Are Unmeasurable

The Objection. "$A(t)$, Φ, and N are vague aggregate variables with no clear operationalization. Without an independent protocol to measure them, the equation is unfalsifiable—you can always claim that Φ must have exceeded N because the system survived."

The Reply. For any specific system, the variables can be operationalized independently of survival. For a cell, $A(t)$ can be measured via membrane potential, intracellular ion concentrations, and ATP/ADP ratios. Φ can be measured by the rate of ATP consumption devoted to repair and transport. N can be measured by the rate of membrane leakage, protein denaturation, and oxidative damage. These are measurable quantities, and the framework provides a template for defining them in any substrate. The crucial point is that the operationalization must be *independent* of the survival outcome—we measure integrity and work rates, then predict persistence, rather than inferring fitness from survival after the fact. That independence breaks the tautology and makes the equation falsifiable.

29.3.3 Some Systems Lack a Clear Threshold

The Objection. "For many complex systems—ecosystems, societies, minds—there is no sharp A_{min} threshold. They degrade gradually without a single point of dissolution. Your equation's binary alive/dead dichotomy is too simplistic to capture these cases."

The Reply. The framework acknowledges that A_{min} can be a region rather than a single value, and that death can be a process rather than an instantaneous event. The binary condition $A(t) > A_{\text{min}}$ is a useful idealization for systems with well-defined boundaries. For systems without a clear physical boundary—like ecosystems—the framework's application requires first identifying the relevant organizational boundary. If no boundary can be identified, the system is not a single agent but a network of agents with nested and overlapping *Hormē*. The equation then applies to each agent boundary individually. For an ecosystem, the dissolution of a species is a sharp event, even if the ecosystem as a whole degrades gradually. The framework provides a multi-scale diagnostic, not a single all-or-nothing judgment.

29.4 Objections to the Epistemology of Asymmetric Risk

Chapter 18: The Epistemology of Asymmetric Risk argued that the prey animal's hair-trigger fear response is not a cognitive error but an operationally true map of the causal stakes: a false negative costs everything, a false positive costs little, so the optimal model heavily weights ambiguous stimuli as threats. Three objections challenge the coherence and implications of this claim.

29.4.1 The Over-Inclusiveness Objection

The Objection. "If any fear that can be rationalized by asymmetric risk is declared 'true,' the framework immunizes itself against all criticism. Phobias, superstitions, paranoia—everything becomes a 'true map' as long as it once contributed to survival. This is unfalsifiable."

The Reply. The framework provides a precise, falsifiable criterion. A map is operationally true only if its net effect on $A(t)$ is positive in the organism's actual environment. A phobia of harmless spiders that causes a person to burn down their apartment imposes costs that far exceed the ancestral benefit of spider avoidance, and it decreases the boundary margin. The map that over-weights spider-danger is true only within a range; when the costs of the false positive approach the costs of the false negative (or exceed them entirely, as in the apartment-burning case), the map is no longer accurate. The framework distinguishes adaptive caution from maladaptive excess by measuring actual outcomes, not by taking the organism's word for it. A map that reliably decreases $A(t)$ is a false map by definition, and the criterion is open to empirical test.

29.4.2 The Modern Mismatch Objection

The Objection. "Many fears are clearly irrational in the modern environment, even if they were rational ancestrally. The person who fears flying but not driving is making a massive calibration error; flying is far safer. The framework seems to excuse this by appealing to ancestral dangers, but that simply means the map is outdated—false for the current world."

The Reply. The framework agrees completely. A map that was causally accurate for an ancestral environment but is miscalibrated for the present one *is* a false map relative to current causal structure. It will degrade $A(t)$ if acted upon—avoiding airplanes while routinely driving imposes a higher risk of death, and if the person's goal includes survival, this behavior is a causal error. The framework's role is to explain *why* such mismatches persist: the *Kriterion* of emotional learning is slow to recalibrate because it evolved in an environment where the cost of a single false negative was infinite, and it has not yet had sufficient selective pressure to adjust to the modern risk landscape. The framework does not excuse the error; it locates its origin and provides a diagnostic for identifying when an ancestrally rational heuristic has become maladaptive.

29.4.3 The Truth-Deflation Objection

The Objection. "By redefining truth as 'what conduces to persistence,' the framework robs truth of its ordinary meaning. When I say 'the shadow is not a predator,' I am making a claim about the world that is simply true, regardless of my survival stakes. You are conflating pragmatic utility with genuine truth."

The Reply. The framework does not deny the ordinary, propositional concept of truth. The statement "the shadow is a predator" is false, full stop. But the organism is not asserting that proposition. Its internal model is not a sentence; it is a sensor-*Organon* coupling that triggers flight. That coupling is not evaluated by whether it corresponds to the visual identity of the stimulus; it is evaluated by whether it correctly tracks the causal risk structure—the expected impact on $A(t)$ of the two possible responses under uncertainty. The framework explicitly distinguishes between *representational truth* (accuracy of a proposition or a photograph) and *operational truth* (accuracy of a causal coupling relative to boundary stakes). Both are real; neither reduces to the other. The claim is that, for the purpose of understanding life and agency, operational truth is fundamental, and representational truth is a derived luxury that complex agents can afford when the stakes are low and time is abundant. Calling the rabbit's startle "true" in the operational sense does not deflate ordinary truth; it expands the concept to its proper domain of embodied agency.

29.5 Objections to the Ontological Mandate of Minimal Determinism

Chapter 19: The Ontological Mandate of Minimal Determinism argued that the very existence of active engines proves that the universe possesses minimal determinism—enough causal regularity for actions to have reliable consequences. Five objections challenge this claim from the directions of fundamental physics, dynamical systems theory, epistemology, scope, and metaphysics.

29.5.1 The Quantum Indeterminacy Objection

The Objection. "Quantum mechanics has shown that the universe is fundamentally indeterministic at the micro level. Events at the quantum scale are irreducibly random. If the world is random at its base, then minimal determinism is false, and agency is an illusion built on macroscopic averaging."

The Reply. Minimal determinism does not require determinism at the micro level. It requires only that there are *some* stable regularities at the scale at which agents act. Quantum indeterminacy is entirely compatible with macroscopic causal stability—that is why chemistry and biology work. The probabilities that quantum mechanics specifies are themselves lawlike and stable. An electron has a definite probability of tunneling through a barrier, and that probability is the same every time the same conditions obtain. That is a regularity. Agents exploit statistical regularities just as effectively as deterministic ones; a bacterium swimming up a nutrient gradient benefits from the fact that diffusion is statistically predictable, even if individual molecular collisions are random. Minimal determinism is the condition that there exists *any* level of description at which actions have reliable consequences. Quantum mechanics satisfies that condition.

29.5.2 The Chaotic Systems Objection

The Objection. "Many natural systems are chaotic, meaning that arbitrarily small differences in initial conditions lead to exponentially diverging outcomes. In such systems, long-term prediction is impossible. If much of the world is chaotic, then the regularities agents rely on are vanishingly narrow, and agency is effectively impossible outside tightly controlled domains."

The Reply. Chaos does not eliminate regularity; it limits the timescale of prediction. A chaotic system is still governed by deterministic or probabilistic laws; its short-term behavior is highly predictable, even if its long-term behavior is not. Agents adapt to this by operating on the relevant timescales. The rabbit does not need to predict the fox's position a year hence; it needs to predict the next ten seconds. Moreover, life often exploits the very fact of chaos. A fleeing prey animal generates chaotic motion by zigzagging, making its trajectory unpredictable to the predator. The predator, in turn, evolves strategies for dealing with this unpredictability—pouncing at close range, using predictive heuristics.

The cat-and-mouse game occurs entirely within a regime of minimal determinism at short timescales. Chaos does not refute the mandate; it reveals the timescale over which the mandate must hold for agency to operate.

29.5.3 The Humean Skepticism Objection

The Objection. "Hume argued that we have no rational justification for believing in causal regularities. The fact that the sun has risen every day does not prove it will rise tomorrow. Induction cannot be justified. Therefore, the 'mandate of minimal determinism' is a statement of faith, not a proof."[4]

The Reply. The framework does not attempt to justify induction from abstract first principles. It accepts Hume's point that the Principle of the Uniformity of Nature cannot be deductively proven. Instead, it exposes a performative contradiction at the heart of the skeptic's objection: the physical act of making the objection requires the very causal regularity the objection denies.

If the world were not inductively tractable—if past regularities provided no operational map for future states—the boundary of any active engine would instantly dissolve. The skeptic must burn energy, fire neurons, and execute directed mechanical work just to vocalize their doubt. Therefore, if the objector exists to make the claim, the premise of causal regularity is already proven. The agent does not need a disembodied, mathematical guarantee that the future will resemble the past. It needs only that its *Kriterion* achieves a net-positive bias toward persistence. The literal, physical existence of the skeptic is the definitive refutation of radical skepticism.

This is not a pragmatic shrug; it is a formal transcendental deduction. It uses strict logic to demonstrate that the causal regularity of the universe is the necessary precondition for the empirical fact of biological persistence. We can map this proof directly:

The Transcendental Proof of Causal Reality

1. **The Degradation Axiom:** In a non-inert universe, physical boundaries continuously degrade as a matter of mathematical necessity.

[4]David Hume, *An Enquiry Concerning Human Understanding* (Oxford University Press, 1748), Sec. IV

2. **The *Kybernēsis* Axiom:** To persist against this degradation, an entity must be able to bias future states by tracking and exploiting actual causal outcomes.
3. **The Performative Axiom:** Life exists and actively persists. (To deny this requires an active engine to expend directed energy to voice the denial, resulting in a performative contradiction).
4. **The Ontological Deduction:** Therefore, the structure of reality is causally regular, at least to the extent required for a living boundary to track and exploit it.
5. **The Epistemological Deduction:** Therefore, the internal model of a living system (*the Kriterion*) tracks actual causal effects, not mere psychological conjectures.

The Local Order Objection

The Objection. "The argument only proves that the local environment of any agent must be minimally deterministic. It says nothing about the universe as a whole. There could be vast realms of pure chaos elsewhere, and agency would still be possible in stable pockets. The mandate is parochial, not cosmic."

The Reply. The framework accepts this. Minimal determinism need not be a global property of the universe. It need only hold in the domain accessible to agents—the sphere of possible interaction. For any given agent, that is the relevant universe. The argument does not claim that the entire cosmos is ordered, only that wherever agency exists, the local causal structure is sufficiently regular. That is enough to ground science, learning, and the navigability of the world as encountered. Moreover, if agents can expand their domain—by building instruments, traveling, or theorizing—they may discover that order extends further than their immediate environment. The history of science suggests this is so. But the logical mandate requires only local order; global order is an empirical discovery.

29.5.4 The "Determinism Eliminates Agency" Objection

The Objection. "If the universe is governed by deterministic laws—if every event is the inevitable outcome of prior states—then agents cannot 'redirect' anything. The causal flow is a single, unalterable rail, and any feeling of choice or intervention is an illusion. The bacterium does not *redirect* the flow; it merely follows its predetermined track. Agency, as the framework describes it, is an empty word."

The Reply. This objection rests on a fundamental category error: it confuses the rules of interaction with a script of destiny.

The laws of reality—the *Logos*—dictate how mass, energy, and information interact. For blind, inert matter, the outcome is highly regular because blind things follow the laws blindly. A falling rock does not measure its trajectory or expend internal energy to alter its course; it simply yields to the immediate forces acting upon it.

But there is no rule in the book of physical laws that says, "matter cannot sense and act." In fact, the laws explicitly permit configurations of matter that measure their environment and spend stored energy to change their state. When life enters the picture, the dynamic fundamentally shifts. An active engine does not break the laws of physics; it uses them. It leverages the reliable regularity of the laws to evaluate conditions and deploy energy to steer.

The key to this architecture is that the laws govern how things change from one moment to the next, but they do not contain a pre-written conclusion. The organism does not need to violate reality to escape a predetermined fate; it uses the regularities of reality to actively write a local future that would not have existed without its intervention.

The objection therefore conflates "law-governed" with "locked in." Lawful regularity is not a prison; it is the reliable grammar the *Kriterion* uses to survive. The blind flow of the universe is regular, but once an entity capable of sensing and spending energy emerges, the future is forged in the act of navigation. The rock yields to the laws; the bacterium navigates through them. Neither breaks them. Only one is an agent.[5]

29.6 Objections to the Nested Architecture of *Hormē*

Chapter 20: Nested Architecture and Conflict Resolution argued that complex agents are composed of nested *Hormic* loops—cells within tissues

[5]The complete architectural derivation of this causal decoupling is beyond the scope of this volume; it forms the central subject of the forthcoming book: *The Living Refutation.*

within organs within organisms, each with its own *Kriterion* and reference state—and that internal conflict is the inevitable condition of such architectures. Three objections challenge the coherence and implications of this nested model.

29.6.1 The Homunculus Objection (Infinite Regress)

The Objection. "You describe a hierarchy of *Kriteria* up to the *Nous.* But a *Nous* that monitors and overrides lower loops is just another *Kriterion*; this leads to an infinite regress of ever-higher *Kriteria.* Who governs the *Nous*? If nothing does, the hierarchy is incomplete. If something does, you need a *Nous*-governor, and then a governor of that governor, ad infinitum."

The Reply. The regress stops at the highest boundary that is practically maintained—the whole organism. The *Nous* is the highest-order *Kriterion* within the individual; there is no further *internal* boundary above it. The environment serves as the ultimate arbiter: if the *Nous*'s decisions consistently degrade the organism dies. The *Nous* can reflect on its own processes—meta-cognition—but this is not a separate governor standing above it; it is the *Nous* applying its modeling capacity to itself. The loop closes with survival. The regress is not infinite; it is bounded by the outermost boundary of the agent in question, and its adequacy is tested by the only tribunal that matters: continued existence.

29.6.2 The Tyranny of the Whole Objection

The Objection. "If the *Nous* always trumps lower goods, the framework justifies a mind-over-body tyranny. The *Orexis* and *Thymos* become mere subjects to be suppressed. But what about the wisdom of the gut? The intuition, the embodied intelligence, the emotional signals that the *Nous* cannot fully articulate? Your hierarchy sounds like a recipe for authoritarianism within the self."

The Reply. The framework does not assert that the *Nous* should always suppress lower strata. The *Nous* depends on the *Orexis* for life, the *Thymos* for social navigation, and the *Logistikon* for efficient action. A wise *Nous* listens to the signals from below—gut feelings, social intuitions, practiced skills—as crucial data that inform judgment. The goal is integration, not domination. The *Nous*'s job is to arbitrate when lower loops conflict or when a lower loop's output would produce a clear net harm to the whole;

it steps in to prevent disaster, not to micromanage well-functioning subsystems. The ideal of *Dikaiosynē*—functional harmony—is precisely the condition in which each stratum contributes its proper function without being crushed or ignored. This is integration, not tyranny.

29.6.3 The Dissolution Objection

The Objection. "If the higher-order governor fails, you claim the system dissolves. But many systems—addicts, crumbling empires, declining ecosystems—persist for a long time in a degraded state. They don't simply vanish at a single point. Your binary alive/dead threshold is too sharp to capture real-world decline."

The Reply. The framework acknowledges that dissolution can be a prolonged process. $A(t)$ can drift downward over years, crossing through stages of dysfunction before finally dropping below $A_{\min}$. The addict's life may slowly unravel; the empire may decay over centuries. The key diagnostic point is not the precise moment of final collapse but the *direction* of $A(t)$: a sustained negative trend signals a failure of governance, even if the system lingers above the threshold for some time. The binary condition $A(t) > A_{\min}$ is a useful idealization for systems with well-defined boundaries; for systems with diffuse or layered boundaries, the framework provides a multi-scale diagnostic that tracks integrity at each level. The sharp threshold is the limit case; the general concept is the gradient of decline.

29.7 The Fourth Wall Holds

The fourth wall remains intact. The framework does not smuggle in teleology, nor does it rely on ungrounded empirical leaps to deduce evolution. Its mathematics are operational, its epistemology rescues the rationality of the frightened animal, its ontology securely demands a navigable universe, and its psychology maps the inevitable conflicts of a nested mind without collapsing into infinite regress.

For over a century, a single argument has stood as the ultimate gatekeeper against any attempt to ground the "good" in the natural world: G. E. Moore's Open Question. If "good" is merely the physical fulfillment of a boundary's persistence, the critic asks, can we not still intelligibly ask, "But is that actually good?" If this objection holds, the normative heart of the derivation is severed from its physical body. To secure the framework permanently, we must face this historic challenge directly. We turn now to the final philosophical battle—and the transcendental closure that wins it.

Chapter 30

The Last Stand: The Moorean Open Question and the Transcendental Closure

We have arrived at the final and most durable objection. The critic has tested the boundary, the inevitability of disruption, the *Kybernēsis* architecture, and the constitutive imperative. Each has held. Now the critic brings forth a weapon that has outlasted a century of rebuttals: G. E. Moore's Open Question Argument. Moore's point is simple, elegant, and devastating to any naturalistic ethics. For any proposed definition of "good"—pleasure, utility, evolutionary fitness, *Hormē*-fulfillment—one can always ask, "But is that thing *actually* good?" The question seems open. The fact that it can be asked without obvious absurdity suggests that "good" is not identical to any natural property.

If this objection stands, the entire normative arc of the derivation—Propositions 10 and 11, the objectivity of value, the truth-goodness isomorphism—collapses. The framework might still describe a physical architecture, but it could not claim to have *grounded value.* This chapter therefore takes the Moorean challenge as the ultimate test. It does not dismiss the question as misguided; it answers it directly, using a transcendental argument that reveals the Moorean's own question is only possible from within the commitment it claims to question.

30.1 Moore's Challenge and Its Hidden Assumption

Moore's argument, presented in *Principia Ethica* (1903), is an appeal to semantic intuition.[1] Take any proposed naturalistic definition of "good"—say, "good means pleasure." Now ask: "I know this is pleasant, but is it *good?*" If the definition were correct, the question would be as empty as "I know he's an unmarried man, but is he a bachelor?" But it is not empty.

[1] G. E. Moore, *Principia Ethica* (Cambridge University Press, 1903), Ch. I, §13

A thoughtful person can intelligibly ask whether pleasure is good, and can even answer "no." Therefore, "good" does not simply mean "pleasant." The same reasoning, Moore insisted, applies to any natural property. The question always remains open, so goodness cannot be identical to anything in the natural world.

The derivation claims that "good" for a system is the fulfillment of its *Hormē*—the directed work that maintains its boundary against dissolution. The Moorean now steps forward and asks: "I know that X fulfills the system's *Hormē*, but is X *actually* good?"

For the test to have its full force, the questioner must be able to consider the candidate property from a position of normative neutrality. Moore assumes that such neutrality is always available—that a reflective agent can step back from any natural property and ask, without prior commitment, whether it is truly good. This assumption is the argument's enabling condition. If the evaluator were already constitutively committed to treating a particular property as good, the test would be biased. And it is precisely this assumption, the *Hormē* framework contends, that cannot hold when the property under scrutiny is the very condition of the questioner's own existence as an evaluating being.

30.1.1 The Transcendental Structure of the Question

The reply to Moore rests on a recognition that has been building throughout this book: the person who asks "Is *Hormē*-fulfillment good?" is not a disembodied intellect. He is a living, bounded, far-from-equilibrium system. His body is, in the very moment of asking, continuously performing the directed work of *Hormē*—maintaining ion gradients, repairing membranes, regulating temperature, pumping blood. These processes are not optional. They are the constitutive striving of his own organism, and they are what keep his brain alive long enough to formulate the question.

This yields a transcendental argument, one that asks not what the words mean but what must be true for the question to be asked at all.[2] It can be stated in three steps:

[2]The full argument is developed technically in Eli Adam Deutscher, *The Transcendental Closure of the Open Question: Why Moore's Challenge Cannot Touch a Hormē-Based Definition of Good*, Neo-Pre-Platonic Press, 2026, `https://neopreplatonic.com/papers/moore/`; the present chapter gives its essential structure in the idiom of the book.

1. **Normative questioning is an activity performed by living systems.** There is no asking "What is good?" without a biological agent whose neural and metabolic machinery makes the asking possible.
2. **Any such system continues to exist only if the directed work of boundary maintenance—*Hormē*—is continuously being performed.** Cessation of that work is dissolution; there is no neutral pause.
3. **Therefore, *Hormē*-fulfillment is a condition of the possibility of normative questioning.** The act of asking is physically impossible without it.

From this, a remarkable conclusion follows. The questioner cannot coherently doubt whether *Hormē*-fulfillment is good, because the act of doubting is itself an instance of the very standard being doubted. The doubt would require the questioner to step outside the conditions of his own existence as a questioner—and that is metaphysically impossible. The question is answered before it is formulated, not by a linguistic rule but by the physics of being alive.

30.1.2 The Illusion of Openness

Why, then, does the question *feel* open? Why can a thoughtful person sincerely ask, "I understand that my body strives to persist, but is that striving *really* good?" The answer lies in the mind's capacity for abstraction. I can represent my own metabolic processes as objects of thought while bracketing the fact that those processes are, right now, sustaining the thought. The mind's ability to treat its own preconditions as optional objects of reflection is the cognitive engine that makes long-term planning possible. But it also generates a characteristic illusion: the sense that I am a neutral observer of my own nature, that I can evaluate the goodness of my striving from a standpoint untainted by it.

This felt gap is an artifact of abstraction, not a disclosure of normative structure. It is like the illusion that one can doubt one's own existence while doing the doubting. Descartes saw that "I think, therefore I am" is not an inference but a recognition: the performance of doubt affirms the existence of the doubter. The *Hormē* framework extends this insight: "I evaluate, therefore I am already treating persistence as an end." The openness of the question is an illusion that dissolves the moment the questioner's own embodiment is brought back into view.

30.1.3 The Performative Contradiction

The illusion becomes a performative contradiction when the doubt is pressed sincerely. Consider an agent who says:

> "I know that my body—its cells, its organs, its neural circuits—is continuously performing the work of boundary maintenance. I know that this work treats persistence as an end at every level of my biology. I know that my capacity to reflect on these facts depends on that work continuing uninterrupted. But I am not convinced that this striving is *good*. The question remains open."

What should we make of this statement? The agent professes normative agnosticism while describing a physical architecture that, in the same moment, treats *Hormē*-fulfillment as good in the only way a physical system can treat anything as good—by acting to maintain the boundary that keeps the agent alive. The cells are not waiting for the verdict of reflection before they continue metabolizing. The heart is not pausing until the agent decides whether persistence is worth it.

If the agent were *truly* agnostic about *Hormē*-fulfillment—if the doubt were not merely verbal but existentially enacted—the system would simply dissolve. There is no intermediate state between treating persistence as an end and ceasing to exist. The sincere doubt cannot be lived. The agent can say the words, but he cannot mean them with his whole being. his biology will not permit it.

This is the signature of a performative contradiction. It does not require a logical inconsistency within the sentence. It requires only that the act of utterance presuppose what the utterance denies. And that condition is met.

30.1.4 Not Analytic, Not Synthetic: Constitutive Identity

One might still ask: what *kind* of definition is the *Hormē* framework offering? It is not an analytic definition—"good" is not synonymous with "*Hormē*-fulfillment" in the way "bachelor" is synonymous with "unmarried man." A competent speaker can intelligibly differentiate the two concepts, and the framework has no interest in denying this. Nor is it simply a synthetic property identity, like "water is H_2O"—a discovery that the two terms happen to refer to the same stuff, which might have turned out otherwise. The identity here is deeper. It is *constitutive*: *Hormē*-fulfillment is the property that makes the practice of normative judgment possible. It picks out the condition that constitutes the questioner as the kind of being who can ask.

An analogy helps. Suppose a physiologist defines health as "the proper functioning of an organism's physiological systems." A skeptic could ask, "But is proper physiological functioning *actually* healthy?" The question feels open only if one imagines "health" as a mysterious, non-natural property floating free of the body. But in the context of a living organism, health just is the state in which its systems are functioning within their viable parameters. The skeptic who continues to ask "but is it *really* health?" is not raising a genuine inquiry; he is expressing a semantic restlessness that no definition can satisfy, because he demands that a functional concept be validated by a non-functional intuition. The *Hormē* definition of good works the same way. It identifies the operational core of goodness—the condition under which any living system's striving succeeds—and it notes that this core is presupposed by the very act of asking.

30.2 Objections and the Nihilist's Defeat

Three objections deserve a hearing.

The nihilist says: "I recognize that my body strives, but I do not care. Nothing is good. Value is a projection, and I refuse to project." The reply is that the nihilist's utterance is a performative act produced by a system that, at every sub-personal level, already treats certain outcomes as worth bringing about. The heart pumps; the lungs exchange gas; the neurons fire. The speech act is parasitic on the biology it disavows. The nihilist can utter the words, but he cannot enact the indifference. Genuine, all-things-considered nihilism is physically unlivable.

The reflective endorser, following Christine Korsgaard, argues that raw biological drives become normative only when an agent steps back, reflectively assesses them, and actively identifies with them as reasons.[3] *Hormē*-fulfillment, she suggests, is just another drive awaiting endorsement. The

[3]Christine M. Korsgaard, *The Sources of Normativity* (Cambridge University Press, 1996), pp. 92–93

reply is that the act of reflective endorsement is itself a temporally extended process requiring a persisting reflective self—and the persistence of that self is a boundary-maintenance problem, solved by *Hormē*. Reflective endorsement does not transcend *Hormē*; it is *Hormē* made self-conscious. The Korsgaardian agent who asks "Should I endorse my drive to persist?" can ask it only because she is already, at the level of architecture, endorsing it with every heartbeat.

The Martian evaluator imagines an alien life form whose constitutive imperative is not persistence but something else—say, the maximization of chaos. The Martian asks: "I know X fulfills Earthling *Hormē*, but is X good?" and the question seems open from the Martian's perspective. This objection forces a valuable clarification: the *Hormē* framework defines what is good *for a system*, not what is good from a God's-eye view that transcends all systems. The Martian's question is perfectly intelligible, but it is not Moore's open question. It is a question about inter-system normativity—about how to adjudicate conflicts between agents with different constitutive aims. That is a genuine and difficult problem, and the framework acknowledges it as an open frontier. But it does not reopen the question of what "good for the Earthling" means. Within the Earthling's own architecture, the answer is settled.

30.2.1 The Two-World Hypothesis (The Dualist's Escape)

The Objection. "You have grounded value entirely in the physical mechanics of boundary maintenance in *this* world. But what if true value—the Good, the moral Ought—exists in a transcendent, non-physical realm? A 'second world' of pure Kantian reason? In that case, biological persistence is merely an instrumental chore, and true goodness is completely independent of physical survival. By trapping value in the logic of the boundary, you are blind to the higher world where real normativity resides."

The Reply. The framework does not deny that a complex agent can imagine a second, transcendent world of pure value. What it points out is that *imagining a second world is an energy-intensive cognitive action performed in the first world.*

The philosopher who posits a transcendent realm of pure reason is not a disembodied intellect; they are a bounded, far-from-equilibrium system.

Their speculation is funded by the continuous operation of *Hormē*. The neurons that conceive of the Kantian reason require tightly regulated ion gradients; the lungs that draw breath to argue for Kantian duty require a continuous expenditure of directed work. The very act of theorizing about a realm beyond the physical is itself a physical achievement, dependent at every instant on the successful maintenance of the thinker's own boundary.

To even formulate a theory of a "second world," an agent must already be satisfying the constitutive imperatives of the first world. If the philosopher's contemplation of the transcendent Good causes them to ignore the approaching predator, the temperature dropping below freezing, or the depletion of their own caloric reserves, their boundary dissolves. When the boundary dissolves, the second world vanishes with it—because the second world was always a model constructed by the *Nous* of a living organism, sustained by the metabolic striving of a body that never pauses to ask permission.

Therefore, the dualist's escape fails via the same performative contradiction as the nihilist's denial. The philosopher uses the biological triumph of *Hormē* to fund a cognitive model that denies the primacy of *Hormē*. The "second world" is not a refutation of the physical grounding of value; it is a metabolic luxury afforded by a system whose lower-order loops are so successful that they allow the *Nous* the free time to dream. The only world where value gets argued about is the world where agents must do work to persist. If a transcendent Good exists, it is utterly impotent until an embodied agent metabolizes enough glucose to speak of it.

30.2.2 The Normative Force Objection

The Objection. "Even if *Hormē*-fulfillment is constitutively inescapable—even if I cannot coherently doubt it while living—you have only shown that I am forced to treat it as an end. You have not shown that I *ought* to. A fact about my constitution is still just a fact. You have bridged the Is-Ought gap only by emptying 'ought' of its distinctive normative force."

The Reply. The objection assumes that normativity must come from outside the agent's constitution—from a realm of pure reasons, or a divine command, or a rational principle that binds all possible beings. The *Hormē* framework denies that assumption. The "ought" is not an alien force; it is the internal, constitutive pressure of being the kind of thing one is. A fish ought to swim, not because it has consulted a moral law, but because its body is organized for swimming and it drowns if it does not. A living system ought to strive to persist, not because the universe commands it, but because its architecture is the architecture of striving, and the alternative is dissolution.

The demand for a further, external "ought" beyond this is like demanding that the law of gravity provide a *reason* to fall. The framework does not reject normativity; it relocates it from a mysterious non-natural property to the causal structure of the agent itself. That is not a deflation of "ought"; it is an explanation of where it comes from and why it has the grip on us that it does. The fish does not resent the water; it is the water that makes its swimming possible. We do not resent *Hormē*; we are *Hormē*, and the very resentment would be another expression of it.

30.2.3 The Final Wall Holds

Moore's open question argument has, for over a century, served as the gatekeeper of metaethical naturalism. It appeared to show, with elegant simplicity, that no natural property could ever be identical to goodness. The *Hormē* framework does not dispute the test's validity for pleasure, desire, evolutionary fitness, or any other natural property that a system might or might not treat as an end. What it shows is that the test cannot be applied to the one property that makes the test possible. The evaluator cannot coherently evaluate the condition of his own evaluating.

The closure is not a piece of linguistic legislation. It is a transcendental recognition, performed by every living agent in the act of persisting. The good is not a mystery at the boundary of the natural world. It is what happens at the boundary of a living thing, continuously, without permission, without pause, for as long as the thing exists. To ask whether that is good is to ask whether existence is good. And the asking answers itself.

We have now defended the derivation from its foundational axioms to its highest structural implications. But one final test remains. A universal theory of life must be able to categorize not just what exists, but what *could* exist. It must possess diagnostic power at the absolute frontiers of reality. In the next chapter, we push the framework to its absolute limits, confronting the ambiguous, the engineered, and the alien. We turn, finally, to the final edge.

Chapter 31

The Final Edge: Edge Cases

The derivation is complete. The eleven propositions have been laid out, defended, and tested against the strongest objections in the philosophical repertoire. The framework now stands as a unified, deductive account of life, agency, and value, grounded in the bare logic of the boundary and the inevitability of disruption. But a framework that claims universality must do more than withstand the attacks of the past. It must prove its worth against the unknowns of the future and the edges of the present. It must be able to classify not only the mule and the bacterium, the virus and the flame, but also the entities that do not yet exist—the synthetic organisms being designed in laboratories, the artificial intelligences that may one day claim agency, and the alien life-forms that may be discovered in the oceans of Europa or the clouds of Venus.

This final chapter of the objections section is therefore not a defense against a skeptical critic but an exploration of the framework's diagnostic power at its outermost boundaries. It asks: can the derivation handle cases that are genuinely ambiguous, genuinely novel, genuinely alien? If it can, it earns its claim to be a universal tool of classification and understanding. If it cannot, it reveals its own limits—and limits, honestly acknowledged, are the mark of a mature theory.

31.1 The Synthetic Organism: Life by Design

In 2010, the J. Craig Venter Institute announced the creation of the first synthetic cell, Mycoplasma mycoides JCVI-syn1.0.[1] The organism's

[1]Daniel G. Gibson et al., "Creation of a Bacterial Cell Controlled by a Chemically Synthesized Genome," *Science* 329, no. 5987 (2010): 52–56, `https://doi.org/10.1126/science.1190719`.

genome was chemically synthesized from digital information and transplanted into a recipient cell from which the original DNA had been removed.

Critically, this was not the creation of life ex nihilo, but a genome transplantation—a form of engineered hijacking.[2] The synthetic genome acted, in its initial stages, much like a virus: it leveraged the existing *Hormē* of the recipient cell, using the host's metabolic *Organon* and its established *Kriterion* references to express its own code. However, as the synthetic genome began to direct the synthesis of its own proteins and lipids, it eventually replaced the host's hardware entirely. At that point, the "parasitic" relationship ended, and a new, self-maintaining agent emerged.

The framework has no difficulty with this case. The synthetic cell is a bounded, far-from-equilibrium system that continuously operates a *Kriterion*-action loop for its own boundary maintenance. Its *Hormē* is intrinsic because its *Kriterion* references are now its own membrane integrity and internal homeostasis. The diagnostic question remains: whose boundary integrity defines the *Kriterion*'s reference state? In the Venter cell, the synthetic genome eventually assumes full responsibility for the "must" of the loop.

But suppose the synthetic cell were modified to be a pure tool—a cell that metabolizes only when an external pump supplies a critical nutrient it cannot sense, or divides only when an external signal triggers fission. In that case, the cell's boundary integrity is contingent upon an external agent's *Hormē*. It would be a "derived agent," much like a mitochondrial organelle or a specialized virus, whose existence is a subordinate function of a larger system's *Logos*. The line between an engineered organism and an organic tool is not blurred; it is drawn by the locus of the reference state. Even in minimal genomes, there remains a "basement" of essential genes that cannot be removed because they are the physical custodians of this loop.[3]

[2] Carole Lartigue et al., "Genome Transplantation in Bacteria: Changing One Species to Another," *Science* 317, no. 5838 (2007): 632–38.

[3] Clyde A. Hutchison et al., "Design and Synthesis of a Minimal Bacterial Genome," *Science* 351, no. 6280 (2016): aad6253.

31.2 The Artificial Intelligence: Why the Robot Objection Fails

The question of artificial intelligence has haunted this book from the early chapters. A thermostat, a chess engine, a language model—each performs discrimination, comparison, and action. Are they agents? Are they alive?

These are tools with derived agency. Their *Kriterion* references are externally imposed; their physical persistence depends on external power and maintenance; they do not maintain their own boundaries. A language model running on a server is not an active engine; it is a process on an engine. The server is the active engine, and the server's active maintenance is performed by human engineers. The AI, considered as software, has no *Hormē* of its own.

But now the critic presses further. "What if we build a robot that maintains its own physical integrity—that seeks charging stations when its battery is low, that avoids damage, that repairs its own circuits, that even builds improved versions of itself? Surely at some point it crosses the threshold. It is acting for its own persistence. Is it not alive?"

The answer is categorical: **no**. The robot objection fails forever. And the reason it fails forever is not a failure of imagination about future technology. It is a logical point about the origin of the reference state.

The robot's drive to recharge its battery, to avoid damage, to repair itself—all of this was programmed by human engineers. The reference state—"maintain battery above 20%," "avoid impacts exceeding 5 newtons," "replace worn actuators"—was chosen by a human designer. Even if the robot can modify its own code, the *meta-criterion* by which it decides what constitutes an improvement was supplied externally. The robot's entire loop, however elaborate, executes a goal that was set from outside. It is not its own goal; it is our goal, enacted through machinery.

We can make this vivid. Imagine a robot that behaves indistinguishably from a living animal—it hunts for fuel, flees from threats, seeks shelter, even appears to show distress when damaged. Now suppose its human designer, sitting at a console, changes a single parameter: the robot now values its own destruction. It seeks out cliffs to drive off, it avoids charging stations, it embraces threats. The robot does this because its reference state was not its own; it was a variable in a program we wrote. An organism's reference state—its set-points, its vital parameters—cannot be changed at a console. They are the physical conditions of its own boundary's existence. You cannot reprogram a bacterium to value acid over glucose, because the acid kills it. Its "preference" is not a choice; it is a constitutive condition.

The line is therefore absolute. Derived agency is real agency—the robot genuinely redirects causal flow—but it is agency in service of an external *Hormē*. Intrinsic agency is agency in service of one's own boundary maintenance, where the reference state is generated by the self-organizing, self-maintaining physical dynamics of the system itself. The two are not points on a spectrum; they are distinct kinds, distinguished by the origin of the reference state. A thermostat and a bacterium both compare and act. The thermostat's reference state was set by a human. The bacterium's reference state is the physical attractor of its own far-from-equilibrium organization. That difference is unbridgeable, no matter how complex the thermostat becomes.

One might ask: could an AI *originate* its own reference state from its own boundary condition? If it were embodied in a physical chassis whose integrity was maintained by processes that self-organized rather than being engineered—if its set-points emerged from the physics of its own far-from-equilibrium dynamics, as they do in a cell—then it would be an intrinsic agent. But that would require that the AI not be programmed with its goals at all. Its goals would have to arise from the self-maintaining dynamics of its own physical organization. That is not a computational problem; it is a physical one. It is the problem of abiogenesis, not the problem of software engineering. Until such a self-originating physical system is built, all robots, however sophisticated, are tools.

The robot objection fails forever because clever programming can only elaborate derived agency. It cannot conjure the intrinsic kind. The difference is one of origin, not of complexity.

31.2.1 The Bootstrapping Trap

The critic's last hope is "bootstrapping"—using biological components to create an artificial system (a bio-hybrid computer). But this does not create a new category of agency; it merely shifts the moral landscape.

If we build a computer out of living neurons or synthetic cells that maintain their own boundaries, we have not created "artificial" intrinsic agency. We have simply hijacked existing intrinsic agency. We have yoked the *Hormē* of a living system to the *Logos* of a human designer. This is not a breakthrough in software engineering; it is a problem of bioethics. It is the enslavement of a living thing, forcing its internal loops to serve external ends. The "bio-robot" is an intrinsic agent whose *Hormē* has been redirected, but the source of that agency remains the self-organizing biophysics of the cell, not the cleverness of the engineer.

31.3 The Alien: Life Beyond Carbon

The framework's substrate-neutrality is one of its strongest claims. It says: wherever a bounded, far-from-equilibrium system maintains its boundary through a continuous comparison-action loop, there is life, regardless of chemistry. This prediction will be tested spectacularly if and when we encounter extraterrestrial organisms.

Consider a hypothetical organism on Titan, Saturn's largest moon. The surface consists of lakes of liquid methane and ethane at temperatures of -179°C. If organisms exist there, they cannot be carbon-based in the terrestrial sense; their membranes might be formed of acrylonitrile polymers, and their metabolism could be based on acetylene hydrogenation.[4]

If such organisms maintain a boundary—perhaps an "azotosome" composed of nitrogen-rich molecules[5]—selectively transport materials, and bias outcomes toward their own persistence, they are alive in exactly the framework's sense. The *Kriterion* might be a molecular switch sensitive to acetylene concentration; the *Organon* might be a cilia-like structure that propels the organism. The chemistry is alien; the logic is identical.

This substrate independence suggests that life is not a specific chemical recipe, but a specific physical arrangement—a topological refusal to dissolve that can manifest in regimes far beyond molecular chemistry.

This substrate independence extends to even more extreme environments. Some theoretical models propose plasma-based "life-forms"—coherent,

[4] Christopher P. McKay and Heather D. Smith, "Possibilities for Methanogenic Life in Liquid Methane on the Surface of Titan," *Icarus* 178, no. 1 (2005): 274–76.

[5] James Stevenson et al., "Membrane Alternatives in World-Scale Non-Aqueous Solvents," *Science Advances* 1, no. 1 (2015): e1400067.

self-sustaining magnetic structures in a stellar corona that maintain their shape against the solar wind. These "complex plasmas" can self-organize into helical structures that exhibit life-like behaviors, such as division and competition for ambient plasma.[6]

If such a structure possesses a boundary (a magnetic envelope), harnesses energy to resist dissipation (magnetic reconnection), and exhibits a *Kriterion* loop (adjusting its field configuration based on ambient density), it meets the criteria. It has been demonstrated in laboratory settings that such plasma spheres can possess distinct double-layer boundaries and exhibit self-organizing replication.[7] The framework would classify it as alive. Whether such entities exist is an empirical question; the framework tells us what to look for: the presence of a self-generated reference state directed toward the maintenance of a physical boundary.

The framework provides a diagnostic checklist for exobiologists: does the candidate entity have a boundary (a membrane, a phase-separated compartment)? Does it maintain that boundary against a disruptive environment? Does it discriminate environmental states and couple that discrimination to actions that bias its persistence? If yes to all three, it is alive. No need to search for DNA, ribosomes, or carbon. The framework's categories are universal.

31.4 The Minimum Possible Life: A Deductive Bound

The derivation also allows us to reason about the minimum possible configuration that could constitute life. Could a single molecule be alive? No, because a single molecule cannot maintain a boundary; it is a passive structure that degrades with thermal fluctuations. Could a minimal autocatalytic set enclosed in a self-assembling lipid vesicle be alive? It would have a boundary and a crude metabolism, but the derivation says life also requires a *Kriterion*—a mechanism that discriminates persistence-relevant states and couples that discrimination to action. A bare vesicle that passively leaks and grows is not an active engine; it is a dissipative structure. To be alive, it would need at least one sensor-*Organon* coupling—perhaps an osmotically activated enzyme that synthesizes more lipid when the

[6] V. N. Tsytovich et al., "From Plasma Crystals and Helical Structures Towards Inorganic Living Matter," *New Journal of Physics* 9, no. 8 (2007): 263.

[7] E. Lozneanu and M. Sanduloviciu, "Minimal-Cell System Created in Laboratory by Self-Organization," *Chaos, Solitons & Fractals* 18, no. 2 (2003): 335–43.

membrane is stretched. Such minimal *Kriteria* are already being developed in protocell research, and the derivation predicts that any successful synthetic life-form will need them.

Thus, the framework sets a deductive bound on the minimal complexity of life. It cannot be simpler than a bounded, far-from-equilibrium system with at least one *Kriterion* that modulates directed work. That is a precise, falsifiable prediction for origins-of-life chemistry and synthetic biology.

31.5 Symbiosis and Hybridity: Mixed Architectures

Where the framework does accommodate gradation is not in the intrinsic/derived distinction itself, but in the architecture of nested and hybrid systems. A lichen, a holobiont, a cyborg with an implanted device—these are not ambiguous agents but systems composed of multiple agents with distinct loci of *Hormē*.

The lichen is a symbiosis of fungus and alga. Each partner is an intrinsic agent with its own *Kriterion* loops. The lichen as a whole is a higher-order boundary with emergent collective dynamics, but its agency is an integration of the partners' intrinsic agencies, not a derived agency imposed from without. The cyborg with an insulin pump has an intrinsic agency (the human) that now includes a tool (the pump) within its boundary. The pump's reference state is derived from the human's *Hormē*; it is a tool integrated into an intrinsic architecture. The human remains the intrinsic agent; the pump remains a derived agent; the system is a hybrid, not a new kind of entity with ambiguous status.

The framework handles these cases by analyzing the *locus of each* Kriterion*'s reference.* For any subsystem, we ask: does this *Kriterion* reference its own boundary integrity, or an externally imposed set-point? A system may be a mosaic of intrinsic and derived elements, but the elements themselves are sharply classifiable. The complexity is in the composition, not in a blurring of the fundamental categories.

31.6 The Objection from Unclassifiable Cases

The critic may make one last objection: "Your framework provides a clean classification only if we already know what the system's boundary is, what its *Kriterion* references, and whether its work is directed. But in complex, novel, or poorly understood systems, these features may be impossible to determine. Your framework is unfalsifiable in practice, even if it is crisp in theory."

This is a pragmatic objection, not a logical one. It is true that applying the framework to a genuinely novel system requires empirical investigation and measurement. But that is a virtue, not a flaw. The framework tells us exactly what to look for. If a candidate system is proposed as alive, we must identify its boundary, measure its energy fluxes, determine whether those fluxes are directed toward boundary maintenance, and locate the *Kriterion* that guides them. These are hard measurements, but they are not impossible. The boundary integral provides a formal criterion: if a system persists with $\Phi \leq N$ over time, the framework is falsified for that system. If we cannot measure Φ or N, the question remains open; the framework does not pretend to answer it without data.

In contrast, a cluster-concept definition of life—"it has some of the properties of metabolism, reproduction, evolution"—provides no diagnostic at all for a genuinely novel case. It merely shrugs. The framework may be difficult to apply, but it is applicable. That is the difference between a theory and a list.

31.7 The Unfinished Deduction

The derivation of life, agency, and value presented in this book is complete in its logical spine. But it is not the last word. It is an opening. By grounding the minimal structure of life in the logic of the boundary, it provides a foundation on which a full philosophy of mind, ethics, and politics can be built. The nested architecture of *Hormē* implies a stratified account of the self, as sketched earlier in this book. The objectivity of value for individual systems implies a framework for adjudicating conflicts between systems—a political philosophy of the boundary. The truth-goodness isomorphism implies an epistemology in which vindication and virtue are inseparable. These extensions lie beyond the scope of this book, but they are its natural progeny.

The historical narrative showed that the empirical sciences have been converging on this architecture for a century. The objections have tested its logical strength and found it resilient. The robot objection fails forever, not because we lack imagination, but because the distinction between intrinsic and derived *Hormē* is a logical distinction of origin, not a matter of degree. A tool, however elaborate, enacts an external will. An organism, however simple, enacts its own. That is the line, and no engineering will ever erase it.

The mule, the bacterium, the virus, the AI, and the alien all find their places in the architecture of the boundary. What remains is to show how a human being, the most complex of all known agents, can use that architecture to navigate its own existence with clarity, purpose, and alignment with reality. The final part of this book turns to that navigational protocol—the practical wisdom that flows from the deduction when it is applied to the living of a human life. The Is–Ought gap is closed; the task now is to walk across the bridge.

Part VII

Lived Navigation

Chapter 32

The Larger Architecture: Neo-Pre-Platonic Naturalism

This book has been a work of deduction. It began with a single, undeniable premise—that a determinate entity must possess a boundary—and proceeded, step by logical step, through eleven propositions to the unification of life, agency, and value. The argument has been self-contained; it has relied on no external authorities, no empirical contingencies, and no assumptions beyond the bare preconditions of existence in a non-inert world. If the deduction holds, it stands on its own feet.

But a deduction, however rigorous, does not exhaust the questions that a thinking being can ask. It tells us what life *is*, why agency and value are inherent in its architecture, and how the Is-Ought gap dissolves upon inspection. It does not, by itself, provide a complete account of the cosmos in which life finds itself, the mind that reflects upon it, or the practical methods by which a human agent can align its own models with reality. These questions belong to the larger philosophical system within which the derivation is embedded: Neo-Pre-Platonic Naturalism (NPN).

This chapter provides a concise summary of that system, showing how the eleven propositions occupy the central pillar of a broader framework that spans ontology, epistemology, ethics, and the philosophy of mind. Readers who wish to explore the full architecture are directed to the foundational text, *Neo-Pre-Platonic Naturalism: First Principles*,[1] where the system is laid out in its entirety. The purpose here is to show that the derivation is not an island but the keystone of an arch.

---

[1]Eli Adam Deutscher, *Neo-Pre-Platonic Naturalism: A First-Principles Framework for Reality, Mind, and Knowledge*, First Edition (Neo-Pre-Platonic Press, 2025).

32.1 The Ground: *Archē* and *Apeiron*

NPN begins with a fundamental distinction that echoes Proposition 1 of the derivation, but at a cosmic scale. To exist as a determinate entity requires a boundary. But what is the ultimate context within which all boundaries are drawn? What is the background against which the totality of determinate reality stands out?

The *Archē* (Ἀρχή) is the term for determinate reality as a whole—the totality of all that is, the closed causal system of objective, physical existence. It is the sum of every bounded entity, every interaction, every law, every gradient. The *Archē* is what there is.

But what is there *not*? The derivation's Proposition 1 established that identity requires a contrast. The *Archē*, as the totality of all determinate things, must itself be contrasted with something. That something cannot be another determinate thing, for then it would be part of the *Archē*. It must be the *indeterminate background* against which determinacy itself emerges. NPN calls this the *Apeiron* (Ἄπειρον)—the boundless, the indefinite, the field of pure potential that is not itself a thing but is the condition for things to be.

The *Apeiron* is not a place, a substance, or a god. It is the category for everything that is not part of the *Archē*, and by definition, it is beyond the reach of empirical observation or logical demonstration. We can know that it must exist—because there must be a contrast for determinacy—but we can know nothing specific about it. It is the ultimate epistemological boundary, the permanent mystery that frames all knowledge.

This dual structure—*Archē* and *Apeiron*, the determinate and the indeterminate—is the ground of NPN's ontology. It generalizes the logic of the boundary to the scale of existence itself. Just as an organism is a bounded "this" against a "not-this," so the entire cosmos of determinate things is bounded by the indeterminacy that is not a thing at all.

32.1.1 The Order Within: *Logos, Hylē, Philia,* and *Neikos*

Within the *Archē*, NPN identifies two fundamental principles that structure all existence:

- ***Hylē*** (Ὕλη): the dynamic, undifferentiated physical potential of the *Archē*—the "stuff" capable of taking all forms. It is not matter in the modern sense of discrete particles, but the underlying capacity for manifestation, always already structured by *Logos.*

- ***Logos*** (Λόγος): the inherent, discoverable order that structures *Hylē*—the complete set of laws, patterns, and regularities that govern all behavior within the *Archē.* The *Logos* is what makes minimal determinism possible. It is the reason the rabbit's leap can work, the bacterium's chemoreceptor can track glucose, and the scientist's equations can predict the eclipse.

The *Logos* is not a static blueprint. It is exhaustively expressed through a fundamental polarity: ***Philia*** (attraction, union, cohesion) and ***Neikos*** (separation, division, individuation). Every phenomenon in the *Archē* arises from the interplay of these two forces—the tendency of things to come together and the tendency of things to pull apart. A stable structure, from an atom to a galaxy to a society, is a dynamic equilibrium of *Philia* and *Neikos.* Too much *Philia* and the structure collapses into undifferentiated sameness; too much *Neikos* and it flies apart into chaos. Persistence, as the derivation showed, is always a balancing act.

This polarity provides the cosmic backdrop for the derivation's core insight. The boundary that defines a living system is a local achievement of the same dynamic: *Philia* holds the boundary together, *Neikos* distinguishes it from the environment. The disruption that Proposition 2 identifies as inevitable is the constant pressure of *Neikos* on the boundary; the directed work of *Hormē* is the ongoing effort of *Philia* to maintain cohesion.

32.1.2 The Emergent Mind: *Nous* and *Somatic logos*

If the *Archē* is the totality and the *Logos* is its order, then what is the mind that comes to know that order? NPN calls this the *Nous* (Νοῦς)—the emergent capacity for abstract modeling that allows a subsystem of the *Archē* to represent, predict, and navigate the larger whole. The *Nous* is not a supernatural substance; it is a physical function, realized in the neural architectures of complex organisms, and—in principle—in any substrate capable of running a sufficiently rich *Kriterion*-action loop.

The *Nous* is the highest expression of the cognitive minimum (Proposition 5). The bacterium's chemoreceptor is a minimal *Kriterion*; the human *Nous* is a *Kriterion* scaled to the ability to model the entire cosmos, the distant past, the remote future, and even itself. The truth-goodness isomorphism (Proposition 11) finds its fullest articulation in the *Nous*: the alignment of the internal model with the *Logos* is both the epistemic pursuit of truth and the ethical pursuit of the good.

But the *Nous* does not reason from pure logical first principles in its everyday operation. It evolved in a specific environment, under specific pressures, and it carries the marks of that history. The evolved heuristics it uses—the rules of thumb, the intuitions, the perceptual categories—are what NPN calls the ***Somatic logos*** (Σωματικός λόγος): the "body-logic" hardwired by evolution, which reliably tracks the macroscopic world well enough for survival but can be led astray in novel or abstract domains. The derivation's account of asymmetric risk is a prime example of the *Somatic logos* at work: the rabbit's hair-trigger fear response is not a logical calculation but an evolved heuristic that correctly tracks the causal stakes.

All knowledge, for finite agents, is a product of the *Nous* applying its *Somatic logos* to the *Logos* of the *Archē*. Because the *Somatic logos* is a product of the *Logos* (it evolved within and by the laws of the *Archē*), there is a fundamental alignment between knower and known. The world is navigable. Science is possible. But the alignment is not perfect; the *Somatic logos* can mislead, dogma can harden, and models must be continually tested against the world. This is the epistemic method of conjecture and refutation, formalized in NPN as the Popperian Protocol.

32.1.3 The Stratified Self: *Psyche, Hormē*, and the Four Strata

The stratified architecture of the human mind as a nesting of *Hormic* loops has been introduced earlier in this book. NPN formalizes this as the ***Psyche*** (Ψυχή)—the total psychological boundary of the individual, composed of four interacting strata:

- ***Orexis*** (Ὄρεξις): the instinctual layer of immediate biochemical drives.
- ***Thymos*** (Θυμός): the strategic social layer of emotion, status, and affiliation.
- ***Logistikon*** (Λογιστικόν): the autopilot, the trained habit-system that automatically governs routine action.
- ***Nous*** (Νοῦς): the conscious executive, the reflective capacity for long-term planning and self-modeling.

At the base of all four, and pervading them, is ***Hormē*** (Ὁρμή)—the constitutive striving that the derivation identified as the engine of life itself. In NPN, *Hormē* is the thread that unites the strata: the body strives to maintain homeostasis; the social self strives for belonging and status; the *Logistikon* strives to execute habits efficiently; the *Nous* strives to align its models with reality. The health of the *Psyche*—what the Greeks called *Dikaiosynē* (Δικαιοσύνη), or functional harmony—is the state in which the *Nous* rightly governs the lower strata, and all four are aligned with the *Logos*.

32.1.4 The Objective Value and the Causal Justice

The derivation's Propositions 10 and 11 established that value is objective and that truth and goodness are isomorphic. NPN embeds these claims in two closely related concepts:

- ***Dikē*** (Δίκη): the impersonal, causal justice of the *Archē*. There is no cosmic judge, but there are objective consequences. A model that misaligns with the *Logos* leads to actions that degrade the boundary. The penalty for error is built into the structure of reality. *Dikē* is what happens when the false map meets the real territory.
- ***Eudaimonia*** (εὐδαιμονία): human flourishing, the objective state of a life lived in alignment with the *Logos*. It is not a feeling (though it often produces feelings of fulfillment); it is the condition of a *Psyche* whose strata are in harmony and whose models are accurate enough to sustain that harmony across time.

The navigational imperative that follows from this is: align your models with reality, or suffer the consequences. This is not a moral command but a causal law. It is as impersonal as gravity. And it applies at every level, from the bacterium that must track the chemical gradient to the civilization that must track its resource base and its internal cohesion.

32.1.5 The Derivation Within NPN

The table below maps the eleven propositions onto the First Principles and Theorems of NPN:

Prop.	Content	NPN Principle
1	Boundary Precondition	GZP (General Zero Principle): identity requires delimitation against an indeterminate background
2	Inevitability of Disruption	T7 (Entropic Asymmetry): maintaining pattern requires continuous energy against dissolution
3	Primal Distinction	FP6 (*Hormē*): the constitutive striving distinguishes active engines from passive conduits
4	Directed Work	FP6, T4 (Ethical Isomorphism): directed work is the operational form of *Hormē*
5	Cognitive Minimum	FP9 (*Nous*): the *Kriterion* is the minimal form of the modeling capacity
6	Model-Action Co-Constitution	FP8 (Navigability): model and action are inseparable in any system that navigates the *Archē*
7	Constitutive Imperative	FP6 (*Hormē*): the must of striving is constitutive for far-from-equilibrium systems
8	Continuous Imperative	T7: persistence is a sustained activity against a standing gradient of dissolution
9	Life-Agency Isomorphism	T6: life and minimal agency are isomorphic
10	Objectivity of Value	C4: good is the fulfillment of *Hormē*; value is an objective functional relationship
11	Truth-Goodness Isomorphism	T4: epistemic error and practical failure are functionally identical

The derivation is thus the proof that NPN's First Principles are not arbitrary postulates. They are deductive consequences of the logic of the boundary. Conversely, NPN provides the cosmic and psychological context that frames the derivation within a complete philosophy of reality, mind, and value.

32.1.6 The Unfinished System

This book has been a work of theoretical biology and meta-ethics, but its implications extend into the most practical of human concerns: how to live. The architecture of nested *Hormē* grounds a political philosophy of the boundary—a theory of justice as the functional harmony of nested social agents. The truth-goodness isomorphism grounds an epistemology in which virtue and vindication are inseparable. The stratified *Psyche* grounds a therapeutic practice—a concrete method for diagnosing and resolving the internal conflicts that prevent human beings from flourishing.

These practical extensions are not developed here. They belong to the larger NPN corpus, and specifically to the forthcoming therapeutic series that will provide a step-by-step guide for individuals seeking to align their *Nous* with the *Logos.*[2] The core of that practical method is the Navigator Protocol—an eleven-stage procedure derived from the deductive logic of the boundary, designed to help a finite, model-building agent audit its own models, recalibrate its *Kriteria*, and bring its entire *Psyche* into functional harmony with reality.

The following chapter provides a brief orientation to that protocol. It outlines the stages, the rationale behind each, and the way they collectively constitute a method of navigation that is at once philosophical and deeply personal. The full development of each stage, with exercises, case studies, and diagnostic tools, is the subject of the forthcoming NPN therapeutic series. The present volume, having established the theoretical foundation, now points the way toward the lived practice.

[2] Eli Adam Deutscher, *The Art & Science of Navigation* (Neo-Pre-Platonic Naturalism, 2026).

Chapter 33

The Navigator Protocol: Aligning with Reality

The derivation of life, agency, and value presented in this book is not merely a theoretical achievement. It is a map. And a map implies a journey. The eleven propositions tell us what we are—bounded agents whose constitutive striving for persistence is the root of all value and whose alignment with the causal structure of the world is the root of all truth. But knowing what we are does not, by itself, tell us how to live. A theory of health is not the same as a treatment plan. A chart of the coastline is not the same as a ship under sail.

The NPN framework includes, as its practical arm, a method of navigation: the Navigator Protocol. This protocol is the operational bridge between the logic of the boundary and the lived experience of a self-aware agent. It is a sequence of eleven stages, each corresponding to a transformation in the agent's relationship to its own models, its own *Psyche*, and the reality it must navigate. The protocol is not a set of arbitrary exercises; it is the logical consequence of the architecture the derivation uncovered. If an agent is a *Kybernētēs* maintaining a boundary through continuous comparison and action, then the improvement of that agent's condition—the restoration of flourishing, the resolution of internal conflict, the correction of error—requires a method that systematically audits, recalibrates, and harmonizes the nested *Kriteria* of the *Psyche*.

This chapter provides a concise overview of the Navigator Protocol. It outlines each stage, its functional definition, and its place in the arc of alignment. The full protocol, with detailed instruction, case examples, and therapeutic guidance, is the subject of the forthcoming NPN therapeutic series.[1] The purpose here is to show that the derivation has a destination: it leads not only to understanding but to practice, not only to knowledge but to the transformation of the knower.

33.1 The Logic of the Protocol

Why eleven stages? The number is not arbitrary, but neither is it magical. It reflects the minimal sequence of transformations that a finite agent must undergo to move from a state of unexamined immersion in its models to a state of reflective alignment with reality. The sequence can be understood as the arc of Socratic examination as formalized by the derivation: first, one must become receptive to the possibility that one's models are inadequate; then one must test them; then one must endure their collapse; then one must rebuild on a firmer foundation. Each stage corresponds to a necessary shift in the agent's reference state—a recalibration of a *Kriterion* within the *Psyche*.

The protocol is not a linear recipe that one completes once. It is a spiral. An agent may cycle through these stages repeatedly, at different scales—a single belief, a whole worldview, a relationship, a career, a life. The stages are the grammar of adaptive self-correction, not a fixed ritual.

33.1.1 The Eleven Stages

The following table summarizes the protocol. The Greek names are technical terms within NPN, each with a precise functional meaning that will be elaborated in the therapeutic works.

Stage	Name	Functional Definition
1	**Eukairia** (Εὐκαιρία)	Receptive Moment. The precondition for inquiry. The agent's defensive structures (*Thymos*) are lowered; the executive (*Nous*) is open to the possibility that its models may be flawed.
2	**Elenchus** (Ἔλεγχος)	Diagnostic Test. Socratic cross-examination of belief. The agent applies cognitive *Neikos* to its own models, testing for contradictions, empirical failures, and violations of logic.
3	**Lysis** (Λύσις)	Dissolution. Structural failure of the old model. When a belief fails the *Elenchus*, it collapses. The agent experiences the loss of a reference state that previously guided action.

[1] Eli Adam Deutscher, *The Art & Science of Navigation* (Neo-Pre-Platonic Naturalism, 2026).

Stage	Name	Functional Definition
4	**Exaiphnes** (Ἐξαίφνης)	The Rupture. Sudden meta-cognitive awareness. The agent sees, not just that a model failed, but that it *was* a model—a construct of the *Nous*, not a piece of the world. This is the birth of reflexive self-modeling.
5	**Katharsis** (Κάθαρσις)	Purification. Releasing attachment to the discarded model. The agent loosens the *Thymic* investment in the old belief—the pride, the identity, the comfort it provided—and makes space for reconstruction.
6	**Aporia** (Ἀπορία)	The Void. Conscious ignorance. The agent confronts the unknown without a map. This is the stage of not-knowing, the necessary ground for genuine inquiry. It is uncomfortable but fertile.
7	**Dikē** (Δίκη)	Re-anchoring. Acceptance of constraints. The agent re-grounds itself in the impersonal causal structure of reality. It acknowledges what cannot be wished away—the objective limits that the *Logos* imposes.
8	**Prohairesis** (Προαίρεσις)	Deliberate Choice. Commitment to a new course of engagement. From the cleared ground of *Aporia* and the acceptance of *Dikē*, the agent commits to a mode of action—whether *Demiourgos* (mastery through analysis) or *Genetor* (creation through synthesis).
9	**Energeia** (Ἐνέργεια)	Being-at-Work. Active implementation. The agent deploys the new model into the causal chain, testing it against reality through sustained, directed action.
10	**Phronēsis** (Φρόνησις)	Practical Wisdom. Automated alignment. Through repetition and success, the new model is encoded into the *Logistikon*—the habit system. The agent no longer needs conscious effort to act in alignment; the alignment has become automatic, intuitive.
11	**Eudaimonia** (εὐδαιμονία)	Flourishing. The objective result. Systemic harmony of the *Psyche* with reality. The agent's nested *Kriteria* are in equilibrium; its models are accurate enough for sustained persistence and fulfillment.

33.1.2 The Arc of Transformation

While the protocol can be applied to any domain, its full power emerges when it is applied to the agent's deepest models—the core reference states that define the self, its values, and its relationship to reality. The arc of the protocol mirrors the structure of the derivation itself. It begins with a boundary (the recognition that one's current map is a map, not the territory), moves through disruption (the collapse of inadequate models), and arrives at a new, more accurate alignment that is continuously maintained by the now-recalibrated *Kriterion* loops.

The protocol is not a technique for feeling better, although alignment often produces relief. It is a technique for *being* better—for bringing one's internal models into closer correspondence with the causal structure of the world, and thereby reducing the friction that causes suffering. Suffering, in the NPN framework, is the signal of misalignment. *Eudaimonia* is the condition of aligning. The protocol is the method for moving from the first to the second.

33.1.3 The Two Modes: *Demiourgos* and *Genetor*

At Stage 8, the agent commits to a mode of engagement. NPN identifies two primary modes, corresponding to the fundamental polarity of *Philia* and *Neikos*:

- ***Demiourgos*** (Δημιουργός): the mode of mastery through structural analysis. The agent decomposes the problem, identifies the principles, and perfects the existing form. It is the mode of the engineer, the logician, the critic. It begins with synthesis (seeing the whole) and proceeds by analysis (breaking it down).
- ***Genetor*** (Γενέτωρ): the mode of creation through synthesis. The agent combines disparate elements, generates novel wholes, and brings new possibilities into being. It is the mode of the artist, the entrepreneur, the innovator. It begins with analysis (the parts at hand) and proceeds by synthesis (combining them into something new).

Both modes are necessary, and both are expressions of *Hormē*. The protocol does not prescribe which mode an agent should adopt at a given moment; it provides the self-awareness to choose deliberately, rather than being unconsciously driven by one's *Somatic logos*. The therapeutic works develop this distinction in depth, with diagnostic tools for identifying which mode is called for by a given situation.

33.1.4 The Protocol as Living Deduction

The Navigator Protocol is not a supplement to the derivation; it is the derivation applied to the task of living. The eleven propositions demonstrated that life, agency, and value are logically inseparable. The protocol demonstrates that the *improvement* of life, agency, and value—the movement from dysfunction to flourishing—is equally logical. It is not a matter of chance or grace, but of method.

The forthcoming NPN therapeutic series will unfold each stage in the detail it deserves. It will provide the reader with concrete exercises for eliciting the *Eukairia*, techniques for conducting the *Elenchus* on one's own beliefs, methods for enduring the *Aporia* without premature closure, and practices for stabilizing the *Energeia* until it becomes *Phronēsis*. The series is intended for anyone who has accepted the derivation's conclusions and wishes to know: *What do I do now? How do I live in light of this understanding?*

This book began with the mule—the sterile, stubborn animal that broke the *a posteriori* definitions of life. It ends with the human, the most complex of all agents, who can, through understanding, break the cycle of unexamined striving and choose alignment. The Navigator Protocol is the map for that choice. The therapeutic series is the compass. The journey is yours.

Chapter 34

The Mandate of the Boundary

A living system is not an object that happens to undergo processes. It is a process that maintains the appearance of objecthood.

We began with a mule. Sterile, stubborn, obviously alive, and yet excluded by every definition of life that biology had managed to produce. The mule was not an anomaly; it was a symptom. It revealed that the *a posteriori* method—abstracting common features from observed organisms—was structurally incapable of capturing what life is. The method could only ever tell us what life happens to look like on this planet, under these conditions, made of this chemistry. It could not tell us what life *must* be, anywhere, in any possible world. And because it could not reach the *must*, it could not bridge the chasm between the physical fact of an organism and the objective value of its existence. The Is–Ought gap remained unspanned, and biology and ethics continued their separate, mutually indifferent courses.

This book proposed a different method. Instead of looking outward at the menagerie of living things and trying to abstract what they share, I looked inward—to the bare preconditions of identity. I asked: what must be true for anything to persist as a distinct entity over time? The answer was the boundary. To be a determinate thing is to be separated from everything else, and that separation is not a biological luxury but a logical necessity. From that single, undeniable starting point, the entire architecture of life, agency, and value unfolded, step by logical step, through eleven propositions.

But the boundary is continuously disrupted. In any world that is not perfectly frozen, interactions cross the boundary and degrade it. Faced with this degradation, entities divide exhaustively into two kinds: passive conduits, which are carried along by the causal flow and dissolve when conditions turn hostile; and active engines, which harness energy to redirect the causal flow toward their own persistence. This is the primal distinction—the first fork in the ontological road, and the exact physical threshold where the category of life begins to separate from the category of mere matter.

An active engine, to succeed, must do more than expend energy randomly. Random expenditure accelerates disorder. The work must be directed—it must bias outcomes, on net, toward boundary maintenance. And direction requires discrimination. The engine must be able to tell the difference between conditions that favor its persistence and conditions that threaten it. This is the *Kriterion*: a physical device, as simple as a protein changing shape upon binding a ligand, that registers a difference and couples that difference to action.

Kriterion and action are not separate modules that an organism happens to have. They are co-constitutive. A *Kriterion* that triggers no action is physically irrelevant to persistence; an action uncoupled from a *Kriterion* is blind flailing. The two arise together, or the system dissolves. This is the loop—the continuous *Kybernēsis* cycle of sense, compare, act, and sense again—that is the operational definition of life.

And this loop is not optional. A far-from-equilibrium system that stops running does not merely become "less fit" or "suboptimal." It ceases to exist. The imperative to persist—*Hormē*—is constitutive. It is what it means to be a bounded system over time. The "must" is not a moral command; it is a law of identity, as impersonal as the necessity that a triangle have three angles. But because the system *must* strive, and because its striving defines certain outcomes as preservative and others as terminal, the loop contains within itself an objective standard of value. "Good" is whatever fulfills *Hormē*; "bad" is whatever frustrates it. Value is not projected onto a neutral universe by human minds; it is enacted by every living thing, from the bacterium swimming up a glucose gradient to the human deliberating about a life choice.

This value is system-relative—what is good for the spider is bad for the fly—but relativity does not imply subjectivity. It implies plurality, grounded in the objective persistence conditions of distinct bounded systems. And because the *Kriterion*'s model must track the actual causal structure of reality to be effective, truth and goodness converge. An accurate model is one whose discriminations correspond to the way the world is; a good action is one that maintains the boundary. But the action is generated by the model. If the model is false at the points where the boundary is at stake, the boundary degrades. Epistemic error and practical failure are the exact same misalignment. The pursuit of truth and the pursuit of the good are the same pursuit.

The Is–Ought gap collapses here. Not because the framework performs an illicit deduction from a bare "is" to a moral "ought," but because it recognizes that the "is" of a living agent is not a static fact, but an ongoing process that already contains its own imperative. The organism does not first exist and then decide to strive; its existing *is* its striving. The ought is embedded in the is, not as a logical inference, but as a constitutive identity.

From this synchronic architecture, the diachronic dimension followed naturally. Individual agents are finite; their boundaries eventually dissolve despite their striving. For the organizational pattern of *Hormē* to persist beyond the individual lifespan, it must be replicated. And because the environment changes over time, replication without variation guarantees eventual extinction. The capacity to generate variation is a logical prerequisite for long-term diachronic persistence—not an empirical accident of imperfect copying, but a structural necessity for any lineage that is not betting on eternal stasis. Among variants, those whose *Kriterion-Organon* loops more effectively bias outcomes toward persistence will persist longer and replicate more. Natural selection is not an independent theory grafted onto biology; it is the diachronic expression of *Hormē*, the same imperative operating across generations. Darwinian evolution is a theorem, not a definition.

The objections were met and answered. The boundary withstood the charge of optionality; disruption withstood the dream of the perfectly closed system. The *Kybernēsis* core withstood the accusations of hidden teleology and semantic inflation. The constitutive imperative withstood the conflation with moralism. The objectivity of value withstood Moore's Open Question, not by a linguistic trick, but by a transcendental argument revealing that the questioner's own asking presupposes the answer. The truth-goodness isomorphism withstood the paradox of the successful psychopath, the martyr's death, and the fitness-beats-truth challenge. The robot objection dissolved entirely, because derived agency, however elaborate, remains the execution of an external will, not the self-originating striving of a boundary that must be what it is.

The derivation thus stands as a unified, deductive proof of the minimal structure of life. It provides a falsifiable diagnostic for any candidate system—natural, synthetic, or alien. It resolves the edge cases that shattered *a posteriori* definitions. And it opens onto a larger philosophical landscape, in which the logic of the boundary grounds an entire architecture of ontology, epistemology, ethics, and practical navigation.

34.1 Swimming in *Hormē*

For most of this book, I have spoken of life in the abstract: bacteria, cells, organisms, boundaries. But the point of the derivation has always been to understand ourselves. And here, at the end, I want to speak directly about what it means to be a human being in light of this architecture.

We are swimming in *Hormē*. Every cell in our bodies is continuously performing the directed work of boundary maintenance—pumping ions, repairing membranes, regulating temperature, fighting pathogens. These processes are not optional. They do not pause for reflection. They are the constitutive imperative made flesh, and they run without ceasing from the moment of conception to the moment of death. We do not feel them because they are so successful. The boundary margin—our $A(t)$—is vast, built up over billions of years of evolutionary refinement and sustained by the coordinated activity of trillions of cells. We are the beneficiaries of a biological inheritance so robust that we can forget it entirely and still survive.

This abundance creates a characteristic illusion. Because our lower-order loops are so effective—because our bodies maintain themselves with such reliability—it feels as though the constitutive imperative is optional, as though value is a matter of choice rather than necessity. We can entertain nihilism. We can contemplate suicide. We can argue, over coffee, that nothing really matters and that all value is a mere projection. We can do this precisely because our bodies are, in the exact same moment, treating our persistence as an end with absolute, metabolic commitment. The philosopher who denies the objectivity of value is funded by the very

Hormē whose reality he rejects. The irony is not a refutation of him; it is the condition of his being able to speak.

The same abundance creates a second illusion: that truth and goodness are decoupled. Because we can hold false beliefs without immediate physical consequence—because believing the sky will fall next Tuesday does not kill us—it seems that epistemic error and practical failure are entirely separate domains. But this is an artifact of our margin. The beliefs that feel like "major" errors are, in fact, peripheral. They concern variables that do not impinge directly on the boundary. The high-stakes truths—that gravity pulls downward, that fire burns, that other minds exist and can harm or help us—are so deeply embedded in our architecture that we do not experience them as beliefs at all. We experience them as reality. The *Somatic logos* has already encoded them into the *Logistikon*, the habit-system that governs our automatic interactions with the world. We do not need to consult our *Nous* to know that stepping off a cliff is bad. The knowledge is in the body, in the gut, in the adrenaline that grips us before we even form a conscious thought.

The more basal the mismatch, the higher the impact on survival. If you believe you can fly and act on that belief, the boundary dissolves instantly. *Dikē* is immediate and absolute. If you believe the wrong political theory, the consequences may unfold over decades, mediated by social and institutional layers that buffer the individual from direct feedback. The apparent independence of truth and goodness is purely a function of distance from the boundary. The closer you get to the edge—the thinner your margin, the harsher your environment, the more directly your model must track reality to keep you alive—the more obviously the isomorphism holds. The bacterium swimming up a gradient has no luxury of error. The human with a full stomach and a locked door has immense luxury. Both are governed by the exact same logic; only the timescale differs.

This is why I have insisted, throughout this book, that the architecture of life is the architecture of value. It is not that the bacterium is a quaint metaphor for human ethics. It is that human ethics is the complex elaboration of the same *Kriterion-Organon* loop that steers the bacterium, scaled up to the level of abstract models, social boundaries, and deep-time planning. The *Nous* is not a separate, magical faculty dropped on top of biology; it is the highest expression of the same discrimination-and-action architecture, operating on representations of the distant future, the social self, and the conceptual order. And it remains answerable to the same absolute arbiter: *Dikē*, the impersonal causal consequence of misalignment, enforced by the dissolution of what is false.

34.2 The Task of Navigation

What follows from this for the living of a human life? The derivation tells us what we are: bounded agents whose constitutive striving is the root of all value, whose alignment with reality is the condition of all flourishing, and whose suffering is the signal of misalignment. The Navigator Protocol, outlined in the preceding chapter, tells us what we can do about it. It is the method for bringing our internal models—our beliefs, our habits, our emotional calibrations—into alignment with the causal structure of reality, thereby reducing the friction that causes suffering and the error that causes dissolution. The protocol is the derivation made practical; the map becomes a journey.

The protocol is not a set of arbitrary exercises. It is the logical consequence of the architecture. If the *Psyche* is a stratified hierarchy of *Kriteria*, each with its own reference states, some of which have become decoupled from the global boundary, then the task of self-governance is to audit those reference states, test them against reality, and recalibrate them where they fail. The *Elenchus* is the method of that audit. The *Aporia* is the discomfort of discovering that a model is false. The *Katharsis* is the release of the attachment to it. The *Energeia* is the sustained work of retraining the *Logistikon* until the new alignment becomes automatic—*Phronēsis*, practical wisdom, the state in which the *Kriterion* tracks the *Logos* without conscious effort.

This is not a project one completes. It is a continuous imperative, just as *Hormē* is a continuous imperative. The boundary never stops degrading; the model never stops needing recalibration. The *Nous* that stops questioning its own maps is a *Nous* that has mistaken comfort for truth, and that mistake will, in time, be corrected by *Dikē*—gently, if the margin is wide; harshly, if the margin is thin. The wise agent does not wait for the correction. The wise agent seeks it out.

I have not written this book merely to win a biological argument. I have written it to forge the central pillar—the derivation that grounds life, agency, and value in the unyielding logic of the boundary. The broader philosophical architecture within which this derivation sits—Neo-Pre-Platonic Naturalism—is already laid out. What remains, and what this book makes possible, is the full development of its implications: a political philosophy of the boundary, an epistemology in which vindication and virtue are inseparable, a therapeutic practice grounded in the mechanics of alignment. Those extensions are the work of a lifetime, and they have already begun. But the pillar at the center—the proof that life, agency, and value are a single, indivisible architecture—is complete.

We are not different in kind from the mule, only in the richness of our models and the scope of our boundaries. We too are bounded agents, constitutively driven to persist, objectively anchored to value, answerable to reality for the accuracy of our maps. The difference is that we can know this. We can understand the architecture of our own striving. We can deliberately recalibrate our *Kriteria*, audit our models, and choose alignment over error.

We can navigate.

The Is–Ought gap is closed. The bridge is built. The task now is to walk across it. The logic of the boundary has brought us to the threshold of a practice. The rest is not deduction but action—the continuous, directed work of a human life, striving in the light of its own understanding, refusing to dissolve.

Glossary of Key Terms

This glossary collects the principal technical terms employed throughout this book, both those drawn from the Neo-Pre-Platonic Naturalist (NPN) framework and those introduced in the course of the derivation itself. Entries are given in alphabetical order. For terms with a Greek origin, the anglicized transliteration appears first, followed by the polytonic Greek in parentheses. English technical terms are interleaved alphabetically. Cross-references to the main text and to the relevant Propositions are provided where helpful.

Active Engine. See *Primal Distinction*; Proposition 3 (Chapter 5).

Aisthēsis (Αἴσθησις) → Empeiros (Ἔμπειρος) → Epistēmē (Ἐπιστήμη) The epistemic sequence: raw sensory perception (*Aisthēsis*) gives rise to pattern recognition and experience (*Empeiros*), which, when structured into causal models that successfully predict and guide action, becomes knowledge (*Epistēmē*). *Epistēmē* is understood as a confidence gradient, not absolute certainty (Part VII: The Unfinished Deduction).

Akrasia (ἀκρασία) Weakness of will; the state in which a higher-order *Kriterion* accurately models what conduces to the system's overall persistence, yet a lower-order loop (e.g., a hijacked reward pathway) dominates the *Organon* output, producing action that degrades the macro-boundary. *Akrasia* is not a decoupling of model and action but a civil war between nested *Hormic* loops, and it is objectively bad for the agent because it drives $A(t)$ toward dissolution. Discussed in detail in the context of addiction and self-destructive behavior (Chapter 13, 20, and 28).

Apeiron (Ἄπειρον) The indeterminate, boundless background against which the determinate *Archē* arises. The *Apeiron* is not a thing or a place; it is the category for everything that is not part of the *Archē*. By definition, it is inaccessible to empirical observation or logical demonstration. It is the ultimate epistemological boundary (Part VII: The Unfinished Deduction).

Aporia (Ἀπορία) The state of conscious ignorance; the recognition that one's current models have failed and that one does not yet possess an adequate replacement. Stage 6 of the Navigator Protocol. *Aporia* is the necessary ground for genuine inquiry (Chapter 33).

Archē (Ἀρχή) The totality of determinate reality; the closed, causal system of objective, physical existence. The *Archē* is constituted by *Hylē* (dynamic potentiality) structured by *Logos* (inherent order). It is the sum of all bounded entities, interactions, laws, and gradients. Contrasted with the *Apeiron* (Part VII: The Unfinished Deduction).

Aretē (Ἀρετή) Functional excellence; the capacity to fulfill an entity's inherent *Hormē* within the *Archē*. An organism that effectively maintains its boundary and flourishes possesses *Aretē* (Chapter 33).

Boundary Any physical or organizational separation that distinguishes an inside from an outside and constitutes an entity as a determinate "this" against a "not-this." The boundary is the minimal precondition for identity (Proposition 1). It may be sharp or graded, material or functional, simple or nested (Chapter 3).

Boundary Integral The formal, mathematical expression of the constitutive condition for persistence:

$$A(t) = A_0 + \int_0^t (\Phi(\tau) - N(\tau))\, d\tau$$

where $A(t)$ is the organizational state at time t, A_0 the initial boundary capacity, Φ the rate of directed work, and N the rate of disruptive work. Persistence requires $A(t) > A_{\min}$ at all times (Chapter 17).

Cognitive Minimum The minimal, mechanical operation of comparison required for an active engine to bias outcomes toward persistence. It consists of a sensor, a reference state, and an *Organon* coupling. The cognitive minimum is the root of all higher cognition (Proposition 5; Chapter 7).

Constitutive Imperative. See *Hormē*; Proposition 7 (Chapter 9).

Daimonion (Δαιμόνιον) In the Socratic context, a highly attuned faculty of instinctual logic; a pre-rational attunement to the *Logos* that signals misalignment. It is a cultivated sensitivity of the *Somatic logos* (Part VII: The Unfinished Deduction).

Demiourgos (Δημιουργός) The navigational mode of mastery through structural analysis. The *Demiourgos* begins with synthesis (seeing the whole) and proceeds by analysis (decomposition, perfection of existing forms). It is the mode of the engineer, the logician, the critic. Contrasted with the *Genetor* (Chapter 33).

Derived Agency Agency that enacts an externally imposed reference state. A tool, a machine, or an artificial system whose *Kriterion*'s set-point was set by another agent possesses derived agency. Its *Hormē*, if the term can be applied, is extrinsic. Contrasted with *intrinsic agency* (Chapter 11, Chapter 22).

Diachronic Persistence The persistence of an organizational pattern across generations, beyond the lifespan of any individual physical vehicle. It requires replication with variation and is the ground of natural selection (Chapter 16).

Dikaiosynē (Δικαιοσύνη) Functional harmony; justice in the sense of the right ordering of parts within a whole. In the *Psyche*, the state where the *Nous* rightly governs the lower strata (*Orexis, Thymos, Logistikon*) in alignment with the *Logos*. In a society, the state where collective *Hormē* is coordinated without tyranny or dissolution (Chapter 20, Part VII: The Unfinished Deduction).

Dikē (Δίκη) The impersonal, causal justice of the *Archē*. The objective constraints and consequences inherent in reality's structure. A system that acts on inaccurate models eventually encounters *Dikē* in the form of boundary dissolution. Stage 7 of the Navigator Protocol: re-anchoring oneself in the acceptance of objective constraints (Part VII: The Unfinished Deduction, Chapter 33).

Directed Work Energy expenditure that, on net, biases the causal flow toward the maintenance of the system's own boundary. It need not be perfectly efficient; a net positive bias is sufficient (Proposition 4; Chapter 6).

Disruption The continuous degradation of a boundary by external and internal interactions. Disruption is inevitable in any non-inert world and is the problem that *Hormē* must continuously solve (Proposition 2; Chapter 4).

Doxa (Δόξα) Unjustified belief; opinion that has not been rigorously tested against the *Logos* and may therefore misalign the agent with reality (Part VII: The Unfinished Deduction).

Elenchus (Ἔλεγχος) Socratic cross-examination; the rigorous logical interrogation of a belief or model to test its coherence and expose contradictions. Stage 2 of the Navigator Protocol (Chapter 33).

Energeia (Ἐνέργεια) Being-at-work; the active, sustained implementation of a model into the causal chain. Stage 9 of the Navigator Protocol. The agent moves from planning to doing, testing the new alignment in practice (Chapter 33).

Epistēmē (Ἐπιστήμη). See *Aisthēsis* ⟶ *Empeiros* ⟶ *Epistēmē.*

Eudaimonia (Εὐδαιμονία) Human flourishing; the objective state resulting from a life lived in functional alignment with the *Logos.* It is not a subjective feeling but a systemic condition of harmony within the *Psyche* and between the agent and reality. Stage 11 of the Navigator Protocol (Chapter 33, Part VII: The Unfinished Deduction).

Eukairia (Εὐκαιρία) The receptive moment; the precondition for genuine inquiry. A state in which the *Psyche* is open, the *Thymos* is quieted, and the *Nous* is willing to consider that its models may be flawed. Stage 1 of the Navigator Protocol (Chapter 33).

Exaiphnes (Ἐξαίφνης) The sudden, meta-cognitive awareness that accompanies the dissolution of a model. The agent sees, not just that a belief failed, but that it *was* a model—a construct, not a piece of the world. Stage 4 of the Navigator Protocol (Chapter 33).

Genetor (Γενέτωρ) The navigational mode of creation through synthesis. The *Genetor* begins with analysis (the parts at hand) and proceeds by synthesis (combining them into novel wholes). It is the mode of the artist, the entrepreneur, the innovator. Contrasted with the *Demiourgos* (Chapter 33).

General Zero Principle (GZP) The foundational principle that for anything to possess determinate identity, it must exist within a delimited context set against an indeterminate background. The logical root of the boundary precondition (Chapter 3, Part VII: The Unfinished Deduction).

Hormē (Ὁρμή) The constitutive, non-negotiable imperative of a far-from-equilibrium bounded system to perform the directed work required to maintain its own boundary. *Hormē* is not a psychological urge but a physical mandate. It is the engine that turns being into striving, and it is the foundation of all value (Proposition 7; Chapter 9).

Hylē (Ὕλη) The dynamic, undifferentiated physical potentiality of the *Archē*; the "stuff" capable of taking all forms, always already structured by *Logos* (Part VII: The Unfinished Deduction).

Intrinsic Agency Agency whose *Kriterion* reference is the system's own boundary integrity. The system acts for its own persistence, not for an externally imposed goal. Intrinsic agency is isomorphic with life (Proposition 9; Chapter 11). Contrasted with *derived agency*.

Intrinsic *Hormē*. See *Hormē*; *Intrinsic Agency*.

Katharsis (Κάθαρσις) Purification; the release of emotional and cognitive attachment to a model that has been dissolved. Stage 5 of the Navigator Protocol. The agent lets go of the pride, identity, and comfort invested in the old belief (Chapter 33).

Kriterion (Κριτήριον) The physical standard by which a system judges whether its current state conduces to its persistence. The *Kriterion* measures the present condition against a reference state and registers a discrepancy. It is the operational seat of agency and value—the point where a mere physical difference becomes a *good* or a *bad* for the system. Composed of a sensor, a reference, and a coupling to the *Organon*. Replaces the term "comparator" used in earlier formulations. (Proposition 5; Chapter 7.)

Kybernēsis (Κυβέρνησις) The continuous *act* of steering by which a bounded system maintains its boundary through the *Kriterion-Organon* loop – the process itself, as distinct from the *Kybernētēs* (the agent who steers). *Kybernēsis* denotes the decentralized, self-referential cycle of sensing, comparing, acting, and re-sensing that constitutes the operational life of any active engine. It requires no central controller or homunculus; the loop is its own governor. Where *Kybernētēs* names what a living system *is*, *Kybernēsis* names what it *does*, continuously and without cessation. (Proposition 8; Chapter 10.)

Kybernētēs (κυβερνήτης) See *Kybernēsis.*

Life The continuous, active instantiation of *Hormē*—the capacity to model (sense and compare) and act in order to perform the directed work necessary to maintain a boundary against disruption (Proposition 6; Chapter 8). Life and minimal (intrinsic) agency are isomorphic (Proposition 9).

Life-Agency Isomorphism The theorem that a system is alive if and only if it possesses intrinsic *Hormē*, and possesses intrinsic *Hormē* if and only if it is an agent. Life and agency are the same physical architecture viewed from within and from without (Proposition 9; Chapter 11).

Logistikon (Λογιστικόν) The autopilot; the stratum of the *Psyche* that encodes trained habits and learned automaticities. The *Logistikon* executes routine comparison-action loops without conscious oversight, freeing the *Nous* for higher-level tasks. It is trained by the *Nous* through repetition (Chapter 20, Part VII: The Unfinished Deduction).

Logos (Λόγος) The inherent, discoverable, constitutive order that structures the *Hylē*; the complete set of laws, patterns, and regularities governing all behavior within the *Archē*. The *Logos* is what makes minimal determinism possible and what the *Nous* models (Chapter 16, Part VII: The Unfinished Deduction).

Lysis (Λύσις) Dissolution; the structural failure of a belief or model when it fails the *Elenchus*. Stage 3 of the Navigator Protocol (Chapter 33).

Minimal Agency The baseline operational capacity to redirect causal flow toward the maintenance of one's own boundary. It requires no consciousness, deliberation, or nervous system, only the *Kriterion-Organon* loop (Proposition 9; Chapter 11).

Minimal Determinism The ontological condition that the universe possesses *some* stable causal regularities—enough that actions can have reliable consequences—without which agency would be impossible. It is a transcendental precondition for the existence of any active engine (Chapter 19).

Model, Operational The physical architecture of the *Kriterion*: the structural coupling between sensor states and *Organon* outputs that correlates with persistence-relevant features of the environment. It is not a mental representation but a causal coupling (Proposition 6; Chapter 8).

Model-Action Co-Constitution The principle that the operational model and the action it guides are logically inseparable in any persistent active engine. A model without action is inert; action without a model is undirected flailing (Proposition 6; Chapter 8).

Natural Selection The diachronic expression of *Hormē*: the differential persistence and replication of organizational variants whose comparison-action loops more effectively bias outcomes toward boundary maintenance. It follows deductively from finite lifespans, replication with variation, and the constitutive imperative (Chapter 16).

Navigator (Ναυτικός) The emergent *Nous* functioning as a model-builder and feedback-driven agent within the *Archē* (Chapter 33).

Neikos (Νεῖκος). See *Philia* and *Neikos.*

Nous (Νοῦς) The faculty of consciousness; the emergent capacity for abstract modeling and long-term planning. The *Nous* is the highest stratum of the *Psyche*, responsible for reflective self-modeling, the training of the *Logistikon*, and the governance of the lower strata. It is the seat of *Prohairesis* (deliberate choice) (Chapter 20, Part VII: The Unfinished Deduction).

Orexis (Ὄρεξις) The instinctual stratum of the *Psyche*; the immediate biochemical drives for survival and homeostasis (hunger, thirst, pain avoidance, thermoregulation). The *Orexis* operates largely below conscious awareness and demands immediate satisfaction (Chapter 20, Part VII: The Unfinished Deduction).

Organon (Ὄργανον) The instrument that executes the *Kriterion*'s judgment as action. From Plato's term for the tool or instrument through which the soul acts on the world (*Timaeus* 44d–45a). The flagellum of a bacterium, the muscle of an animal, the hormonal cascade of a cell—all are *Organa*: physical mechanisms through which evaluation becomes causal intervention. The *Organon* without the *Kriterion* is blind motion; the *Kriterion* without the *Organon* is inert judgment. Replaces the term "effector" used in earlier formulations. (Proposition 6; Chapter 8.)

Passive Conduit A bounded entity that does nothing to alter the causal flow toward its own persistence. It is carried along by events and persists only as long as the environment happens not to dissolve it. Contrasted with *active engine* (Proposition 3; Chapter 5).

Philia (Φιλία) and *Neikos* (Νεῖκος) The exhaustive cosmic polarity: *Philia* (attraction, union, cohesion) and *Neikos* (separation, division, individuation). Together they constitute the dynamics of the *Logos*. Persistence of any structure is a dynamic equilibrium between the two (Part VII: The Unfinished Deduction).

Phronēsis (Φρόνησις) Practical wisdom; the state in which a model has been so thoroughly integrated that it operates automatically through the *Logistikon.* Stage 10 of the Navigator Protocol. The agent acts in alignment without conscious effort (Chapter 33).

Physis (Φύσις) Nature; the natural order. The *Archē* understood as a dynamic, emergent process, governed by *Logos* and expressed through the interplay of *Philia* and *Neikos* (Part VII: The Unfinished Deduction).

Prohairesis (Προαίρεσις) Deliberate choice; the conscious commitment to one mode of engagement over another based on a holistic assessment of the *Psyche*'s overall good. Stage 8 of the Navigator Protocol. The *Nous* exercises *Prohairesis* when it selects, for instance, the *Demiourgos* or *Genetor* mode (Chapter 33).

Primal Distinction The exhaustive division of bounded entities into passive conduits (which are carried by the causal flow) and active engines (which redirect it toward their own persistence). The first fork in the ontological road (Proposition 3; Chapter 5).

Psyche (Ψυχή) The stratified structure of the human mind: a layering of the core *Hormē* into *Orexis* (instinct), *Thymos* (social emotion), *Logistikon* (autopilot), and *Nous* (conscious executive). The health of the *Psyche* (*Dikaiosynē*) is the functional harmony of these strata (Chapter 20, Part VII: The Unfinished Deduction).

Reference State The physical parameter against which a *Kriterion* measures current conditions. For intrinsic agents, the reference state is the condition of the system's own boundary integrity. For derived agents, it is an externally imposed goal (Chapter 7, 15).

Somatic logos (Σωματικός λόγος) The evolved heuristic logic of the *Nous*; the "body-logic" hardwired by evolutionary success. The *Somatic logos* includes perceptual categories, intuitions, and innate fears. It reliably tracks macroscopic reality but can be misled in novel or abstract domains (Part VII: The Unfinished Deduction).

Synthesis (Σύνθεσις) The core operation of the *Genetor* mode: combining disparate elements to generate novel wholes (Chapter 33).

Teleology The explanation of systems, structures, or actions by reference to an intended future state, goal, or final cause. Teleological explanations assume that a future outcome can actively pull or direct present physical behavior from ahead, an inversion rejected by this derivation in favor of present structural constraints. Contrasted with *teleonomy* (Chapter 15).

Teleonomy Apparent purpose arising from feedback-stabilized mechanisms shaped by past selection. Teleonomic explanations invoke present physical organization and historical differential persistence, not future goals or final causes. Contrasted with *teleology* (Chapter 15).

Thymos (Θυμός) The strategic social stratum of the *Psyche*; the layer of emotion and motivation that manages in-group/out-group dynamics, status, affiliation, and reputation. It operates through social emotions such as shame, pride, anger, and loyalty (Chapter 20).

Threshold of Dissolution (A_{min}) The critical value of the organizational state variable below which the boundary can no longer be restored by any feasible directed work. Once this threshold is crossed, the system ceases to exist as a distinct entity (Chapter 17).

Truth-Goodness Isomorphism The theorem that for an agent, epistemic error (falsehood) and practical failure (vice) are functionally identical: both are misalignments of the *Kriterion*'s model with the causal structure of reality. The pursuit of truth and the pursuit of the good are the same operation (Proposition 11; Chapter 13).

Umwelt The species-specific perceptual world of an organism; the subset of environmental features to which its *Kriteria* are tuned. The *Umwelt* is the organism's operational model of reality, not a subjective bubble but a selective channel that must track causal truth (Chapter 23).

Frequently Raised Objections Index

This appendix provides a cross-referenced table of the principal objections raised against the derivation. The objections are grouped by the proposition or subject area they target. Page references are dynamic; chapter and section numbers direct the reader to the specific location where each objection is presented in its fullest form and answered.

Objections to the Boundary Precondition (Proposition 1)

Objection	Location
The Flat Denial	Ch. 26, §26.1.1
The boundary/identity distinction is circular	Ch. 26, §26.1.2
Fields and wavefunctions lack sharp boundaries	Ch. 26, §26.1.3
"Boundary" is too vague / elastic to be useful	Ch. 26, §26.1.4
Boundaries are observer-dependent, not objective	Ch. 26, §26.1.5
The holobiont and symbiosis dissolve clear boundaries	Ch. 26, §26.1.6

Objections to the Inevitability of Disruption (Proposition 2)

Objection	Location
Crystals in saturated solution persist without energy	Ch. 26, §26.2.1
A perfectly closed system escapes disruption	Ch. 26, §26.2.2
Self-stabilizing boundaries repair without work	Ch. 26, §26.2.3

Objections to the Primal Distinction (Proposition 3)

Objection	Location
Dissipative structures are active but not alive	Ch. 27, §27.1.1
Machines (solar pumps) are active engines but not alive	Ch. 27, §27.1.2
Flames create convection currents that sustain them	Ch. 27, §27.1.3

Objections to Directed Work (Proposition 4)

Objection	Location
"Directed work" smuggles in teleology (final causation)	Ch. 27, §29.1.1
Spandrels and neutral traits falsify universal directedness	Ch. 27, §27.2.2

Objections to the Cognitive Minimum (Proposition 5)

Objection	Location
Mousetraps and thermostats also "compare"	Ch. 27, §27.3.1
Calling physical discrimination "cognitive" is panpsychism	Ch. 27, §27.3.2
True comparison requires intentionality (aboutness)	Ch. 27, §27.3.3
Calling a *Kriterion*'s differential "good/bad" is projection	Ch. 27, §27.3.5
Where does the reference state come from without prior cognition?	Ch. 29, §29.1.4

Objections to the Constitutive Imperative, *Hormē* (Proposition 7)

Objection	Location
The imperative feels optional (the illusion of option)	Ch. 28, §28.1.1
The "must strive" conflates physical necessity with moral "ought"	Ch. 28, §28.1.2
Suicide violates the constitutive imperative	Ch. 28, §28.1.3
Machines perform directed work but lack intrinsic *Hormē*	Ch. 28, §28.1.4

Objections to the Life–Agency Isomorphism (Proposition 9)

Objection	Location
Viruses redirect causal flow toward replication	Ch. 28, §28.2.1
A chess AI exhibits agency but is not alive	Ch. 28, §28.2.2
Defining life via *Hormē* and agency via *Hormē* is circular	Ch. 28, §28.2.3
Calling a bacterium an "agent" cheapens the term	Ch. 28, §28.2.4
A metabolic mutant cell without sensors would be alive	Ch. 28, §28.2.5

Objections to the Objectivity of Value (Proposition 10)

Objection	Location
Moore's Open Question: "Is *Hormē*-fulfillment *actually* good?"	Ch. 30, §30.1
Value is system-relative and therefore not objective	Ch. 28, §28.3.1
The nihilist can coherently reject the value of their own persistence	Ch. 28, §28.3.2
Deriving "good" from physical facts commits the naturalistic fallacy	Ch. 28, §28.3.3

Objections to the Truth–Goodness Isomorphism (Proposition 11)

Objection	Location
The successful psychopath: cruelty can aid survival	Ch. 28, §28.4.1
Martyrdom: dying for a true cause is an existential failure	Ch. 28, §28.4.2
Addiction and *akrasia*: accurate belief but self-destructive action	Ch. 28, §28.4.3
The fitness-beats-truth paradox (survival selects for useful fictions)	Ch. 28, §28.4.4

Objections to Darwinian Evolution as a Deductive Consequence

Objection	Location
"Survival of the fittest" is a tautology	Ch. 16, §16.3.1
Replication is an empirical premise, not deduced	Ch. 29, §29.2.1
The framework says nothing about the origin of life	Ch. 29, §29.2.2
Lamarckian / epigenetic inheritance undermines selection	Ch. 29, §29.2.3
Neutral evolution and drift are widespread	Ch. 29, §29.2.4

Objections to the Boundary Integral

Objection	Location
The equation is merely a thermodynamic truism	Ch. 29, §29.3.1
The variables are unmeasurable	Ch. 29, §29.3.2
Some systems lack a clear threshold of dissolution	Ch. 29, §29.3.3

Objections to the Epistemology of Asymmetric Risk

Objection	Location
Any fear can be rationalized (the over-inclusiveness objection)	Ch. 29, §29.4.1
Fears are often maladaptive in modern environments	Ch. 29, §29.4.2
Conflating pragmatic utility with genuine truth	Ch. 29, §29.4.3

Objections to Minimal Determinism

Objection	Location
Quantum indeterminacy undermines causal regularity	Ch. 29, §29.5.1
Chaotic systems make prediction impossible	Ch. 29, §29.5.2
Humean skepticism defeats induction	Ch. 29, §29.5.3
Determinism eliminates genuine agency	Ch. 29, §29.5.4

Objections to the Nested Architecture of *Hormē*

Objection	Location
The homunculus objection leads to infinite regress	Ch. 29, §29.6.1
The tyranny of the whole over the parts	Ch. 29, §29.6.2
Systems degrade gradually without a sharp dissolution	Ch. 29, §29.6.3

Bibliography

Adler, Julius. "Chemoreceptors in Bacteria." *Science* 166, no. 3913 (1969): 1588–97.

Aristotle. *De Anima.* Translated by J. A. Smith. Princeton University Press, 1984.

Aristotle. *Politics.* Translated by C. D. C. Reeve. Hackett Publishing, 1998.

Ashby, W. Ross. *Design for a Brain.* Chapman & Hall, 1952.

Bell, Graham. "Evolutionary Rescue and the Limits of Adaptation." *Philosophical Transactions of the Royal Society B* 368, no. 1610 (2013): 20120080. https://doi.org/10.1098/rstb.2012.0080.

Berg, Howard C. *E. Coli in Motion.* Springer, 2004.

Berg, Howard C, and Douglas A Brown. "Chemotaxis in Escherichia Coli Analysed by Three-Dimensional Tracking." *Nature* 239, no. 5374 (1972): 500–504.

Bernard, Claude. *Introduction à l'étude de La Médecine Expérimentale.* Baillière, 1865.

Boltzmann, Ludwig. "Über Die Beziehung Zwischen Dem Zweiten Hauptsatze Der Mechanischen Wärmetheorie Und Der Wahrscheinlichkeitsrechnung Respektive Den Sätzen Über Das Wärmegleichgewicht." *Sitzungsberichte Der Kaiserlichen Akademie Der Wissenschaften. Mathematisch-Naturwissenschaftliche Classe* 76 (1877): 373–435.

Brooks, Rodney A. *Cambrian Intelligence: The Early History of the New AI.* MIT Press, 1999.

Cannon, Walter B. *The Wisdom of the Body.* W. W. Norton & Company, 1932.

Carter, Brandon. "Large Number Coincidences and the Anthropic Principle in Cosmology." In *Confrontation of Cosmological Theories with Observational Data,* edited by M. S. Longair. Reidel, 1974.

Dawkins, Richard. *The Selfish Gene.* Oxford University Press, 1976.

Deacon, Terrence W. *Incomplete Nature: How Mind Emerged from Matter.* W. W. Norton & Company, 2011.

Dennett, Daniel C. *From Bacteria to Bach and Back: The Evolution of Minds.* W. W. Norton & Company, 2017.

Deutscher, Eli Adam. *Neo-Pre-Platonic Naturalism: A First-Principles Framework for Reality, Mind, and Knowledge.* First Edition. Neo-Pre-Platonic Press, 2025.

Deutscher, Eli Adam. *The Art & Science of Navigation.* Neo-Pre-Platonic Naturalism, 2026.

Deutscher, Eli Adam. *The First Cut: How Boundary Logic Derives Physics.* Neo-Pre-Platonic Press, 2026. https://neopreplatonic.com/papers/First_Cut/.

Deutscher, Eli Adam. *The Transcendental Closure of the Open Question: Why Moore's Challenge Cannot Touch a Hormē-Based Definition of Good.* Neo-Pre-Platonic Press, 2026. https://neopreplatonic.com/papers/moore/.

Euclid. *The Thirteen Books of Euclid's Elements.* 2nd ed. Vol. 1, translated by Thomas L. Heath. Dover Publications, 1956.

Friston, Karl. "The Free-Energy Principle: A Unified Brain Theory?" *Nature Reviews Neuroscience* 11, no. 2 (2010): 127–38.

Galperin, Michael Y. "A Census of Membrane-Bound and Intracellular Signal Transduction Proteins in Bacteria: Bacterial IQ, Extroverts and Introverts." *BMC Microbiology* 5 (2005): 35. https://doi.org/10.1186/1471-2180-5-35.

Gánti, Tibor. *The Principles of Life.* Oxford University Press, 2003.

Gibson, Daniel G., John I. Glass, Carole Lartigue, et al. "Creation of a Bacterial Cell Controlled by a Chemically Synthesized Genome." *Science* 329, no. 5987 (2010): 52–56. https://doi.org/10.1126/science.1190719.

Gould, Stephen Jay, and Richard C. Lewontin. "The Spandrels of San Marco and the Panglossian Paradigm: A Critique of the Adaptationist Programme." *Proceedings of the Royal Society of London B* 205, no. 1161 (1979): 581–98. https://doi.org/10.1098/rspb.1979.0086.

Hoffman, Donald D. *The Case Against Reality: Why Evolution Hid the Truth from Our Eyes.* W. W. Norton & Company, 2019.

Hoffman, Donald D., Manish Singh, and Chetan Prakash. "The Interface Theory of Perception." *Psychonomic Bulletin & Review* 22, no. 6 (2015): 1480–506.

Hoyle, Fred, and Chandra Wickramasinghe. *Evolution from Space.* Simon; Schuster, 1981.

Hume, David. *A Treatise of Human Nature.* John Noon, 1739.

Hume, David. *An Enquiry Concerning Human Understanding.* Oxford University Press, 1748.

Hutchison, Clyde A. et al. "Design and Synthesis of a Minimal Bacterial Genome." *Science* 351, no. 6280 (2016): aad6253.

Israelachvili, Jacob N. *Intermolecular and Surface Forces.* 3rd ed. Academic Press, 2011.

Jablonka, Eva, and Marion J. Lamb. *Evolution in Four Dimensions: Genetic, Epigenetic, Behavioral, and Symbolic Variation in the History of Life.* Revised. MIT Press, 2014.

Jonas, Hans. *The Phenomenon of Life: Toward a Philosophical Biology.* Harper & Row, 1966.

Joyce, Gerald F. "The Foreword." In *Origins of Life: The Central Concepts,* edited by David W. Deamer and Gail R. Fleischaker. Jones; Bartlett Publishers, 1994.

Kant, Immanuel. *Critique of Pure Reason.* Translated by Paul Guyer and Allen W. Wood. Cambridge University Press, 1998.

Kauffman, Stuart A. *Investigations.* Oxford University Press, 2000.

Kauffman, Stuart A. *The Origins of Order: Self-Organization and Selection in Evolution.* Oxford University Press, 1993.

Keilin, David. "The Problem of Anabiosis or Cryptobiosis in Living Organisms." *Proceedings of the Royal Society of London. Series B - Biological Sciences* 150, no. 939 (1959): 149–91. `https://doi.org/10.1098/rspb.1959.0013`.

Kirkwood, Thomas B. L. "Evolution of Ageing." *Nature* 270 (1977): 301–4. `https://doi.org/10.1038/270301a0`.

Korsgaard, Christine M. *The Sources of Normativity.* Cambridge University Press, 1996.

Lartigue, Carole et al. "Genome Transplantation in Bacteria: Changing One Species to Another." *Science* 317, no. 5838 (2007): 632–38.

Levin, Michael. "Technological Approach to Mind Every-Where: An Integrated Hierarchy of Emergent Locus of Agency." *Frontiers in Systems Neuroscience* 15 (2021): 652635.

Levin, Michael. "The Computational Boundary of a 'Self': Developmental Bioelectricity as a Bottom-up Mechanism for Biological Systems to

Create Granular Scale-Free Intelligence." *Frontiers in Psychology* 10 (2019): 2688.

Lozneanu, E., and M. Sanduloviciu. "Minimal-Cell System Created in Laboratory by Self-Organization." *Chaos, Solitons & Fractals* 18, no. 2 (2003): 335–43.

Luisi, Pier Luigi. *The Emergence of Life: From Chemical Origins to Synthetic Biology.* Cambridge University Press, 2006.

Maturana, Humberto R., and Francisco J. Varela. *Autopoiesis and Cognition: The Realization of the Living.* D. Reidel, 1980.

Mayr, Ernst. "Teleological and Teleonomic: A New Analysis." *Boston Studies in the Philosophy of Science* 14 (1974): 91–117.

McKay, Christopher P., and Heather D. Smith. "Possibilities for Methanogenic Life in Liquid Methane on the Surface of Titan." *Icarus* 178, no. 1 (2005): 274–76.

Miller, Earl K., and Jonathan D. Cohen. "An Integrative Theory of Prefrontal Cortex Function." *Annual Review of Neuroscience* 24, no. 1 (2001): 167–202. https://doi.org/10.1146/annurev.neuro.24.1.167.

Monod, Jacques. *Chance and Necessity: An Essay on the Natural Philosophy of Modern Biology.* Alfred A. Knopf, 1971.

Moore, G. E. *Principia Ethica.* Cambridge University Press, 1903.

Morowitz, Harold J. *Energy Flow in Biology: Biological Organization as a Problem in Thermal Physics.* Academic Press, 1968.

Nesse, Randolph M. "The Smoke Detector Principle: Natural Selection and the Regulation of Defensive Responses." *Annals of the New York Academy of Sciences* 935, no. 1 (2001): 75–85. https://doi.org/10.1111/j.1749-6632.2001.tb03473.x.

Neumann, John von. *Theory of Self-Reproducing Automata.* Edited by Arthur W. Burks. University of Illinois Press, 1966.

Nicolis, Grégoire, and Ilya Prigogine. *Self-Organization in Nonequilibrium Systems: From Dissipative Structures to Order Through Fluctuations.* Wiley, 1977.

Parmenides. "Fragment 8." In *Die Fragmente Der Vorsokratiker*, 6th ed., edited by Hermann Diels and Walther Kranz. Weidmann, 1951.

Pittendrigh, Colin S. "Adaptation, Natural Selection, and Behavior." In *Behavior and Evolution*, edited by Anne Roe and George Gaylord Simpson. Yale University Press, 1958.

Plato. *Five Dialogues: Euthyphro, Apology, Crito, Meno, Phaedo.* Translated by G. M. A. Grube. Hackett Publishing Company, 1981.

Plato. *Republic.* Edited by C. D. C. Reeve. Translated by G. M. A. Grube. Hackett Publishing Company, 1992.

Plato. *Timaeus.* Translated by Donald J. Zeyl. Hackett Publishing Company, 2000.

Prigogine, Ilya. "Time, Structure, and Fluctuations." *Nobel Lecture in Chemistry*, 1977.

Schrödinger, Erwin. *What Is Life? The Physical Aspect of the Living Cell.* Cambridge University Press, 1944.

Searle, John R. "Minds, Brains, and Programs." *Behavioral and Brain Sciences* 3, no. 3 (1980): 417–24.

Singer, S. J., and Garth L. Nicolson. "The Fluid Mosaic Model of the Structure of Cell Membranes." *Science* 175, no. 4023 (1972): 720–31.

Skinner, B. F. "'Superstition' in the Pigeon." *Journal of Experimental Psychology* 38, no. 2 (1948): 168–72.

Stevenson, James, Paulette Clancy, and Jonathan I. Lunine. "Membrane Alternatives in World-Scale Non-Aqueous Solvents." *Science Advances* 1, no. 1 (2015): e1400067.

Tanford, Charles. *The Hydrophobic Effect: Formation of Micelles and Biological Membranes.* Wiley-Interscience, 1973.

Tononi, Giulio. "Consciousness as Integrated Information: A Provisional Manifesto." *Biological Bulletin* 215, no. 3 (2008): 216–42.

Tsytovich, V. N., G. E. Morfill, V. E. Fortov, N. G. Gusein-Zade, B. A. Klumov, and S. V. Vladimirov. "From Plasma Crystals and Helical Structures Towards Inorganic Living Matter." *New Journal of Physics* 9, no. 8 (2007): 263.

Uexküll, Jakob von. *Streifzüge Durch Die Umwelten von Tieren Und Menschen: Ein Bilderbuch Unsichtbarer Welten.* Julius Springer, 1934.

Varela, Francisco J., Evan Thompson, and Eleanor Rosch. *The Embodied Mind: Cognitive Science and Human Experience.* MIT Press, 1991.

Watson, H. W., and Francis Galton. "On the Probability of the Extinction of Families." *Journal of the Anthropological Institute of Great Britain and Ireland* 4 (1875): 138–44.

Weber, Andreas, and Francisco J. Varela. "Life After Kant: An Introduction to a Naturalistic Teleology." *Phenomenology and the Cognitive Sciences* 1 (2002): 97–125.

Wiener, Norbert. *Cybernetics: Or Control and Communication in the Animal and the Machine.* MIT Press, 1948.

Wittgenstein, Ludwig. *Philosophical Investigations.* Translated by G. E. M. Anscombe. Blackwell, 1953.

Index

About the Author

Eli Adam Deutscher is a philosopher, systems architect, and the founder of Neo-Pre-Platonic Press. His work begins from a functional crisis: the inability of modern frameworks to provide coherent, actionable answers to fundamental questions of knowledge, value, and purpose.

Trained in the rigorous, applied disciplines of engineering and economics at the **Colorado School of Mines** and tested in the complex, strategic arena of consultancy, Deutscher approached philosophy not as a historical debate but as a **first-principles design problem**. *Neo-Pre-Platonic Naturalism* is the result: a complete, deductive meta-structure that regrounds human meaning in the constitutive logic of a lawful cosmos.

He writes for those who require a philosophy that is not merely believed, but **built—and used.**

A Note on the Production

This volume was published by **Neo-Pre-Platonic Press**
(an Imprint of the NPN System) in Spring 2026.

Book Design and Typesetting executed by **Eli Deutscher** using LaTeX and Pandoc.
The production pipeline was done with the assistance of
artificial intelligence. All Ideas are the original work of the Author.
All primary source material and philosophical commentary
are the original work of the author.

Body text is set in Libertinus Serif at 11pt.

The cover was designed by the Author.

www.ingramcontent.com/pod-product-compliance
Lightning Source LLC
Chambersburg PA
CBHW031336230926
51848CB00002B/252
9781971183060